Communications in Computer and Information Science 783

Athula Ginige · Heinrich C. Mayr
Dimitris Plexousakis · Vadim Ermolayev
Mykola Nikitchenko · Grygoriy Zholtkevych
Aleksander Spivakovskiy (Eds.)

Information and Communication Technologies in Education, Research, and Industrial Applications

12th International Conference, ICTERI 2016
Kyiv, Ukraine, June 21–24, 2016
Revised Selected Papers

Springer

Editors
Athula Ginige
Western Sydney University
Penrith, NSW
Australia

Heinrich C. Mayr
Institute of Applied Informatics
Alpen-Adria-Universität Klagenfurt
Klagenfurt
Austria

Dimitris Plexousakis
University of Crete
Heraklion
Greece

Vadim Ermolayev
Zaporozhye National University
Zaporozhye, Zaporiz'ka
Ukraine

Mykola Nikitchenko
Taras Shevchenko National University of Kyiv
Kiev
Ukraine

Grygoriy Zholtkevych
V.N. Karazin Kharkiv National University
Kharkiv
Ukraine

Aleksander Spivakovskiy
Kherson State University
Kherson
Ukraine

ISSN 1865-0929 ISSN 1865-0937 (electronic)
Communications in Computer and Information Science
ISBN 978-3-319-69964-6 ISBN 978-3-319-69965-3 (eBook)
https://doi.org/10.1007/978-3-319-69965-3

Library of Congress Control Number: 2017957679

Printed on acid-free paper

This Springer imprint is published by Springer Nature
The registered company is Springer International Publishing AG
The registered company address is: Gewerbestrasse 11, 6330 Cham, Switzerland

Preface

This volume contains a number of selected and extended contributions to ICTERI 2016, the 12th International Conference on Information and Communication Technologies (ICT) in Education, Research, and Industrial Applications: Integration, Harmonization, and Knowledge Transfer.

The conference was held in Kiev, Ukraine, during June 21–24, 2016, with a focus on research advances in ICT, business/academic applications of ICT, and design and deployment of ICT Infrastructures.

ICTERI 2016 continued the tradition of hosting co-located events, this year by offering three workshops and a PhD Mentors Panel. The workshops addressed: (1) long-standing research aspects of reliability modeling and assessment, (2) cross-disciplinary issues in the use of information technology in economics and finance, and (3) new uses of information technology for life-long learning. The PhD Mentors Panel provided the opportunity to PhD candidates to listen to and discuss promising topics for research offered by several renowned experts.

As in previous years, the ICTERI 2016 proceedings have been published as a CEUR-WS volume (http://ceur-ws.org/Vol-1614/), containing 62 papers selected from a total of 122 submissions. Of these papers, the 18 best were chosen by the program and workshop chairs to be submitted in substantially extended and revised versions for the proceedings volume. Out of these, 16 were resubmitted. Again, these papers were reviewed by at least two experts regarding scientific and technical quality, anticipated reader interest, and coverage of the conference scope. Finally, the Proceedings Committee selected the ten most mature and interesting papers for publication after further revision. The acceptance rate thus is 8.2% regarding the overall number of ICTERI 2016 submissions.

The papers in this volume focus on architecture, augmented reality, case-based reasoning, multi-state systems, cloud computing, scalable parallelism, artificial neural networks, malicious code and intrusion detection, deterministic and stochastic models, didactic models in practice, research-based education and mobile learning environments, and teaching strategies.

In their invited paper, Sandra Stinčić Clarke, John Davies, and Mike Fisher explain the role of an information broker in an open ecosystem, address the challenge of data interoperability in the IoT context and describe the Hypercat standard. This standard allows for a uniform access to distributed data repositories (data hubs) analogously to standard Web protocols and formats like HTTPS, JSON, and REST. Several examples of IoT applications that use data from Hypercat-enabled data hubs are presented.

Mykola Tkachuk et al. present a three-level architecting approach to adaptive resource management in mobile augmented reality systems (MARS). This approach proposes an ontology of adaptive MARS resources on the conceptual level, a generic algorithmic model on the logical level, and a reference software architecture on the physical level.

Myroslav Komar et al. discuss the integration of artificial immune systems (AIS) and artificial neural networks (ANN) as a basis for intelligent cyber defense systems. The idea is to detect network attacks and malicious code based on AIS principles and detectors that have an ANN structure.

Elena Zaitseva et al. present an approach that uses ordered fuzzy decision trees to overcome the problem of constructing a system's structure function in the case of incomplete information.

Eugene Tulika et al. describe the use of rewrite-rules for transforming legacy Fortran applications to be executed in the cloud. Resource allocation is optimized by adopting service choreography.

Gregory Zholtkevych et al. present an approach to reveal relationships between components of natural systems with feedback. Two models are discussed that allow one to determine the direction of pairwise relationships in the deterministic case and the direction and strength of relationships in the stochastic case.

Leo van Moergestel et al. address the process of redesigning a computer science (CS) curriculum and introducing blended learning in a CS educational program. The successes of the new program as well as the problems encountered are discussed.

Yuriy Kondratenko et al. discuss the correlation between research-based education, government priorities, and research funding with a special emphasis on the role of ICT in the education of engineering students. As an example, the use of modeling methods for the implementation of prosthesis and robotics research projects is presented.

Mariya Shyshkina describes a cloud-based learning environment and proposes some indicators for quality evaluation.

Nataliya Osipova et al. address the systematic use of mobile learning technologies for facilitating learning as well as ensuring accessibility, equity, individualization, and flexibility. As an example, a model for learning English phonetics is proposed.

Finally, Nataliya Kushnir et al. report on experiences made when preparing school teachers and university lecturers to use ICT in the educational process. The approach applied was that of master classes, which comes with the advantage of being short term and allowing for productivity at work.

This volume would not have materialized without the support of many people. First, we are very grateful to all the authors for their continuous commitment and intensive work. Second, we would like to thank the Program Committee members and additional reviewers for providing timely and thorough assessments. Furthermore, we would like to thank all the people who contributed to the organization of ICTERI 2016. Without their efforts there would have been no substance for this volume.

June 2017

Athula Ginige
Heinrich C. Mayr
Dimitris Plexousakis
Mykola Nikitchenko
Vadim Ermolayev
Grygoriy Zholtkevych
Aleksander Spivakovskiy

Organization

General Chair

Aleksander Spivakovsky	Verkhovna Rada of Ukraine, Kherson State University, Ukraine

Steering Committee

Vadim Ermolayev	Zaporizhzhya National University, Ukraine
Heinrich C. Mayr	Alpen-Adria-Universität Klagenfurt, Austria
Mykola Nikitchenko	Taras Shevchenko National University of Kiev, Ukraine
Aleksander Spivakovsky	Verkhovna Rada of Ukraine, Kherson State University, Ukraine
Mikhail Zavileysky	DataArt, Russian Federation
Grygoriy Zholtkevych	V.N. Karazin Kharkiv National University, Ukraine

Program Chairs

Athula Ginige	Western Sydney University, Australia
Heinrich C. Mayr	Alpen-Adria-Universität Klagenfurt, Austria
Dimitris Plexousakis	Institute of Computer Science, FORTH, Greece

Workshop and Tutorial Chairs

Vadim Ermolayev	Zaporizhzhya National University, Ukraine
Grygoriy Zholtkevych	V.N. Karazin Kharkiv National University, Ukraine

IT Talks Chairs

Aleksander Spivakovsky	Verkhovna Rada of Ukraine, Kherson State University, Ukraine
Mikhail Zavileysky	DataArt, Russian Federation

Local Organization Chairs

Anatoly Anisimov	Taras Shevchenko National University of Kiev, Ukraine
Mykola Nikitchenko	Taras Shevchenko National University of Kiev, Ukraine
Volodymyr Shevchenko	Taras Shevchenko National University of Kiev, Ukraine

Publicity Chair

Nataliya Kushnir	Kherson State University, Ukraine

Web Chair

Eugene Alferov	German Climate Computing Center, Germany

Program Committee

Costin Badica	University of Craiova, Romania
Nick Bassiliades	Aristotle University of Thessaloniki, Greece
Sotiris Batsakis	University of Huddersfield, UK
Antonis Bikakis	University College London, UK
Anatoliy Doroshenko	National University of Technology KPI, Ukraine
Vadim Ermolayev	Zaporizhzhya National University, Ukraine
David Esteban	TECHFORCE, Spain
Aditya Ghose	University of Wollongong, Australia
Athula Ginige	University of Western Sydney, Australia
Brian Hainey	Glasgow Caledonian University, UK
Sungkook Han	Wonkwang University, Korea
Ievgen Ivanov	Taras Shevchenko National University of Kiev, Ukraine
Dorota Jelonek	Czestochowa University of Technology, Ukraine
Jan Jürjens	Fraunhofer ISST and University of Koblenz-Landau, Germany
Natalya Keberle	Zaporizhzhya National University, Ukraine
Vitaliy Kobets	Kherson State University, Ukraine
Evgeniy Lavrov	Sumy State University, Ukraine
Vladimir A. Shekhovtsov	Alpen-Adria-Universität Klagenfurt, Austria
Claudia Steinberger	Alpen-Adria-Universität Klagenfurt, Austria
Martin Strecker	Université de Toulouse, France
Julius Stuller	ICS CAS, Prague, Czechia
Ilias Tachmazidis	University of Huddersfield, UK
Mauro Vallati	University of Huddersfield, UK
Leo Van Moergestel	Utrecht University of Applied Sciences, The Netherlands
Sergiy Vilkomir	East Carolina University, USA
Paul Warren	Knowledge Media Institute, Open University, UK
Rong Zhang	Nishinippon Institute of Technology, Japan
Grygoriy Zholtkevych	V.N. Karazin Kharkiv National University, Ukraine

Sponsors

Oleksandr Spivakovsky's Educational Foundation (OSEF)	http://spivakovsky.fund/
DataArt	www.dataart.com
Taras Shevchenko National University of Kiev	http://www.knu.ua/en
Logicify	http://logicify.com/

Contents

Invited Paper

The Internet of Things – Technical Challenges for Interoperability

Sandra Stinčić Clarke(✉), John Davies, and Mike Fisher

BT Group Plc., Ipswich, UK
{sandra.stincic, john.nj.davies, mike.fisher}@bt.com

Abstract. In this paper, we look at the various players in the Internet of Things (IoT) and explain the critical role of an information broker in an open ecosystem. We look at the challenge of data interoperability in the IoT context and describe the Hypercat standard; a specification for representing and exposing Internet of Things data catalogues to improve data discoverability and interoperability. The central idea is to enable distributed data repositories (data hubs) to be used jointly by applications by making it possible to query their catalogues in a uniform machine-readable format. This enables the creation of "knowledge graphs" of available datasets across multiple hubs that applications can exploit and query to identify and access the data they need, whatever the data hub in which they are held. This is achieved by employing the same principles on which linked data and the web are built: data accessible through standard web protocols and formats (HTTPS, JSON, and REST); the identification of resources through URIs; and the establishment of common, shared semantics for the descriptors of datasets. We exemplify by way of several demonstrations of IoT applications which use data from Hypercat-enabled data hubs.

Keywords: Internet of things · Information broker · Data interoperability · Distributed data repository · Data discovery · Data cataloguing

1 Introduction

The Internet of Things (IoT) is anticipated to deliver significant innovation in many different areas, with increasing number of applications appearing in the areas such future cities, transport, health and social care, manufacturing, and agriculture [1]. Sensors can be deployed at low cost to measure and monitor to a far greater extent than has been possible before. From a technology point of view, there is little fundamentally new in IoT – it is all about working at a different scale (more devices, more data, more scope for automation), and the potential to share information much more widely, driven by ever decreasing component cost and device miniaturisation. Stakeholders increasingly recognise the potential value in opening up data resources so that they can be exploited more fully. The ability to achieve more effective communications with people wherever they are adds an extra dimension to the potential applications that can be built. There is increasing activity and interest in exploring this rich set of possibilities from around the world. However, the IoT exhibits considerable diversity and complexity which means that significant barriers to innovation remain and present us with a

A. Ginige et al. (Eds.): ICTERI 2016, CCIS 783, pp. 3–13, 2017.
https://doi.org/10.1007/978-3-319-69965-3_1

number of important new challenges. Early IoT implementations focus predominately on sensing and these will be discussed within this paper. We envisage, as the IoT matures, more actuation solutions will be available.

At the highest level, many of the IoT applications being considered appear similar – involving the collection of information from a range of sensors and other sources, interpreting this in a specific context, and then making better decisions which improve behaviour compared with what was done before. There are two important classes of sensor-based IoT application – those which aim to monitor and respond to time-sensitive conditions, and those which collect data over a period of time for analysis. In either case, much of the time and effort involved in building applications can be spent on activities which appear generic. It would be much easier to build a new application in an environment where these generic problems have already been solved by others, with robust solutions available to all. This is what we mean by an IoT ecosystem [2].

This paper explores certain parts of an IoT ecosystem, with specific focus on the role of an information broker in an open ecosystem and practical issues associated with interoperability of the IoT ecosystems, with real life examples of some of the UK cities.

2 Background

2.1 IoT Ecosystems

An IoT ecosystem consists of a number of independent stakeholders sharing a common interest in particular kinds of information, and obtaining benefit from participation. When we say information, we use that as a generic term. Information might come from sensors; it can define physical artefacts or even include aspects of the physical world. Benefit from participation in the IoT ecosystem can be achieved as a commercial provider of information or analytic services, as an application developer, a platform provider, or end user, for example.

Use of shared services and facilities generally involves a compromise – typically giving up some level of direct control in return for reduced costs. In the case of today's global communications networks (including the Internet), the case for common services is very strong. The potential for an IoT ecosystem to enable innovation seems clear, but all participants need to have confidence in the value proposition and be convinced that it meets their needs. If this is not the case then a sustainable ecosystem will not be possible. Areas of concern will include security and trust, respect for personal and commercial rights, dependability, performance, the ability to comply with legal and regulatory obligations, and cost. Predictability, simplicity and flexibility are particularly important characteristics [3].

Additionally, the potential advantages of information sharing are accompanied by real concerns over security and privacy.

2.2 IoT Opportunities for Cities

If we observe a natural ecosystem – a city – the opportunities to explore the benefits that an IoT ecosystem could bring are easy to envisage. Within transport, there are opportunities to reduce traffic congestion, plan journeys more efficiently, reduce accidents and generally improve other services that involve transport and traffic. Those implementations have a positive effect on issues such as air quality, and benefits can be achieved in other areas, such as in optimisation of waste disposal. IoT solutions have potential to be used to enable assisted living at home, or to reduce energy and/or water use. Some combination of many such initiatives can enable sustainable economic growth of a previously constrained city.

2.3 Participants in an IoT Ecosystem

Stakeholders in an IoT ecosystem may take one or more roles. In our work, we distinguish information providers, application developers, analytic service providers, platform providers, and users of information and applications. Additionally, both information providers and users can be either individual or organisational. These roles are shown in the value chain of Fig. 1 and are described in more detail below.

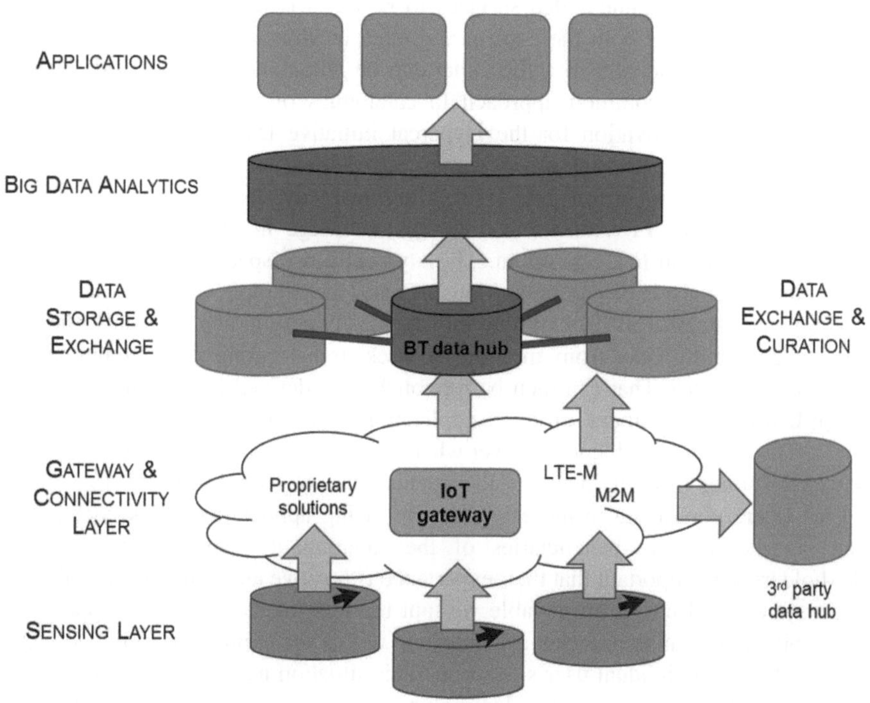

Fig. 1. IoT ecosystem value chain

Information Providers in IoT are often owners of sensor deployments. The primary purpose of their sensors may be for their own use but they may choose to make some of their data available to others, either on a commercial basis, to meet their obligations (particularly for public sector organisations), or for the benefit of society.

Other types of information (other than sensor data) might be available and also of value in IoT applications – for example contextual information (e.g. geographical, administrative, etc.) or information about more complex situations (e.g. traffic incidents, sporting fixtures, etc.) can help to make sense of sensor data.

Information providers should be able to easily publish data resources and advertise their availability via an easily accessible catalogue (such as those using the Hypercat specification [4]) so that potential users can independently discover and assess their utility. Making data available must not imply relinquishing ownership rights so the information provider needs the ability to define access controls as well as terms and conditions for use of their data.

Platform Providers have a key enabling role in the IoT ecosystem. They do not directly provide information or build applications but support stakeholders in other roles by providing a set of common services that all can use, allowing them to focus on their core concerns and accelerating innovation in the ecosystem. They may provide computing and storage infrastructure, offered for example as virtualised cloud services. More specific to IoT ecosystems, platform providers will offer the facility for information resources and value-added services to be advertised to potential users. Catalogues need to describe both the content and other attributes (quality measures, terms and conditions etc.), ideally in a form that can be automatically processed. There is obvious benefit in a common approach to catalogues of information and services, which is a major motivation for the Hypercat initiative [5]. Platform services also include facilities for collecting, transforming, distributing and storing data from many independent and heterogeneous sources in a uniform way.

Analytic Service Providers have started to emerge in the last few years, and naturally fit within an IoT ecosystem. They typically use specialist software tools to enable efficient analysis of particular kinds of data, for example very large datasets or high-frequency streams of data. While not necessarily adding any new data, they add value by combining data from multiple sources, transforming or applying various analytical techniques. That can then be republished as derived information.

Application Developers produce applications typically using a range of data sources. They should be able to discover what data resources are available to them, and assess which ones meet the needs of the applications they want to build.

End Users participate in the ecosystem by using applications made available to them. As the ultimate beneficiaries of the functionality provided by the other stake-holders, it is important that their experience is positive and delivers real value. An IoT ecosystem will not be sustainable without the trust of its end users. Applications may involve information associated with users' use of the application being collected, for example each individual user's location. This situation needs to be addressed with care, particularly where personally identifiable or potentially sensitive information may be involved [6]. Open engagement with end users which ensures they are properly informed and included in the ecosystem is essential.

3 Overview

3.1 Information Brokerage and Data Hubs

Projects such as MK:Smart [7] and CityVerve [8] aim, through real world case studies and applications, to illustrate the value of making information more freely available by improving current ways of working and enabling innovation. Demonstrating good practice, the benefits of making the best use of all available information, and enabling new modes of cooperation between organisations are important contributions which are replicable in other situations.

We anticipate that there will be many IoT ecosystems, each focused on particular kinds of information. For example, they may be based on a constrained geographical area, as in the CityVerve's Oxford Road Corridor, a 2 km long road; or on a city itself, as in the MK:Smart project and city of Milton Keynes. The scope could be administrative (e.g. associated with a city, region or nation) or environmental, such as a river, where flooding and drought can have significant impact on communities. Alternatively ecosystems may naturally develop around concepts such as health and wellbeing, or associated with a global supply chain, rather than a location. As well as the information content itself, individual IoT ecosystems may have specific requirements on security, reliability and latency, particularly as IoT systems develop beyond sensors to include actuators which can directly affect the physical world.

Value analysis across the IoT Ecosystem (Fig. 2) developed within the STRIDE project (one of eight InnovateUK funded IoT clusters [9]) in 2014, showcases interests and involvements of each of the participants within the ecosystem. It also stresses the need for an integrator – a platform provider – capable of supporting the needs of all interested parties.

This diversity of players will clearly lead to distinct and varied technical solutions with a risk that opportunities for sharing potentially valuable information are lost. To avoid this, there is a need for consistency of approach and interoperability. While it is entirely reasonable for an individual or organisation to regard some or all of its data as private or confidential and to be unwilling to make it available to others, unnecessary technical barriers should be eliminated. Interoperability between ecosystems – based, for example, on interoperability between data hubs – is therefore required.

3.2 Hypercat

Hypercat represents a pragmatic starting point to solving the issues of managing multiple data sources, aggregated into multiple data hubs, through linked data and semantic web approaches [10]. Hypercat allows a server (IoT data hub) to provide a set of resources to a client, each with a set of metadata annotations.

Hypercat started as a way of achieving interoperability between multiple data hubs in different domains, between the eight IoT clusters [9].

Hypercat is a specification for representing and exposing IoT catalogues using web technologies (including URIs, HTTP, JSON etc.) to make it easier to identify and discover data resources. If each data hub provides a consistent and uniform machine-readable catalogue, it is possible to create a view of resources across a set of hubs, allowing applications to find relevant data wherever it is held. The Hypercat specification is a pragmatic starting point, solving some important issues facing developers of IoT applications.

While delivering a common web API and file format, Hypercat has a small set of mandatory core metadata (see Fig. 3) but can easily link to domain-specific vocabularies allowing implementers freedom to use any set of annotations to suit their needs.

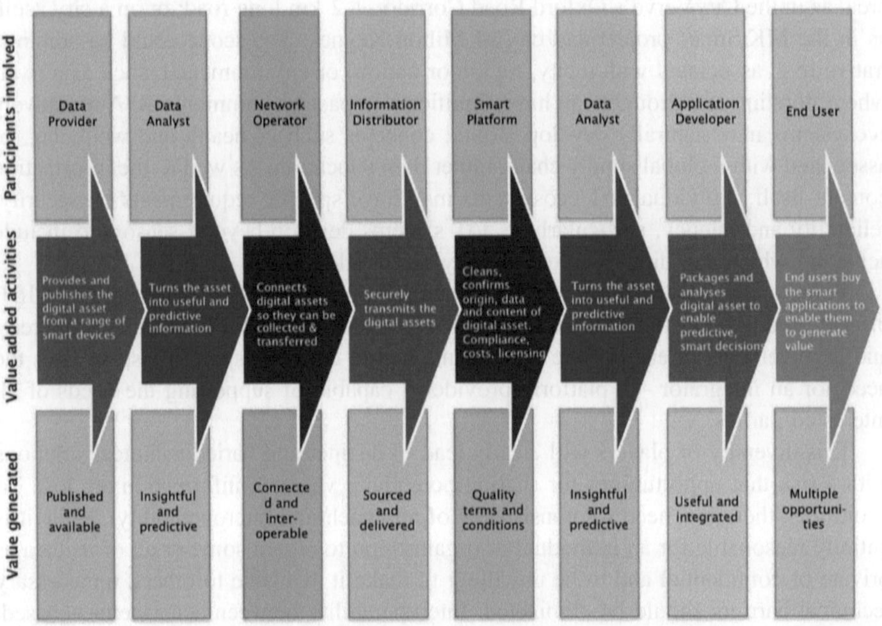

Fig. 2. Value analysis across the IoT Ecosystem

Each of the data resources in the catalogue is annotated with metadata (in a way of RDF-like triples). Potential extensions are needed to utilise the benefits of semantic interoperability, as described in [10]. Extending the core Hypercat with semantic description provides an early indication of benefits of querying distributed data repositories. That extension is enabling querying of the data hub's resources via SPARQL endpoint and creating a complex query that can include external data sets available elsewhere via LOD (Linked Open Data). The newly formed federated queries extends further the data discovery functionality over different internal and external SPARQL endpoints.

```
{
 "catalogue-metadata":[
  {
   "rel":"urn:X-hypercat:rels:isContentType",
   "val":"application/vnd.hypercat.catalogue+json"
  },
  {
   "rel":"urn:X-hypercat:rels:hasDescription:en",
   "val":"Minimal empty catalogue"
  }
 ],
 "items":[]
}
```


Fig. 3. Hypercat minimal empty catalogue

4 Findings and Discussion

4.1 Near Future for Cities

An increasing number of cities will face the need to expose their digital assets and, eventually, implement IoT (or IoT like) solutions to reduce the cost and improve efficiency of managing their infrastructure and/or operations. Those initiatives will open up large amounts of real time, near real time, and historical datasets and allow the opportunities for data aggregation and data analytics. It is of utmost importance to lower the barriers to participation for data providers and application developers (including the ease of data maintenance, avoidance of data duplication, and consideration of security and data privacy restrictions). In any implementation, the notion of avoiding data silos should be kept in mind – both logically and architecturally.

The MK:Smart project explored some of the practical issues associated with creating an IoT ecosystem, focused on the city of Milton Keynes. The project is a broad collaboration engaging a variety of different stakeholders with the aim of solving some real world problems in a sustainable way, considering technical, commercial, organisational, and user experience issues.

The CityVerve project has the aim to create a 'platform of platforms', a flexible, scalable combination of multiple independent platforms containing a secure catalogue of data that be used for discovery and development of novel applications in response to the city's evolving requirements. This will lead the way in identifying next set of interoperability challenges between hubs of different providers and implementations.

4.2 Data Hubs

Within complex IoT Ecosystems, as a composition of e.g. the city, organisations operating within the city, and citizens interacting with them, it is very likely that large amounts of data will be generated daily. The data will comprise from sensor generated data (e.g. temperature or footfall), events (e.g. incidents on the local roads or a theatre show), and a wide (still being classified) selection of contextual information (e.g. geospatial location of the bus stops, paths used by pedestrians, or types of business models for bicycles available for rental). That data might be available from different providers, under different terms and conditions, in different formats, and on different data hubs.

Unless that data is accessible in a more machine readable way, effective use of it will become virtually impossible – simply creating data silos of the IoT kind. Enabling interoperability and data discovery by utilising a format such as Hypercat, for cataloguing IoT resources, provides a solution for this. Having each data hub enriched with a Hypercat data catalogue enables those distributed data repositories to be used jointly by applications by making it possible to query their catalogues in a uniform machine-readable format. This enables the creation of "knowledge graphs" of available datasets across multiple hubs that applications can exploit and query to identify and access the data they need, whatever the data hub in which they are held. This is achieved through employing the same principles on which linked data and the web are built: data accessible through standard web protocols and formats; the identification of resources through URIs; and the establishment of common, shared semantics for the descriptors of datasets.

Additionally, being able to access data from any of the data hubs in a uniform manner – preferably by using same or similar API – has the potential to foster faster innovation. Rapid development of proofs of concept (example of Hypercat4All [5] City Dashboard shown in Fig. 4) is enabled by that uniform exposure. That capability provides opportunity to quickly gauge potential of one's application and lowers the risk of the development cost.

Fig. 4. Increased speed in the prototype creation

4.3 Interoperability

Based on [1], utilising IoT technologies and linking the physical and digital worlds could generate up to $11.1 trillion a year in economic value by 2025. Achieving this kind of impact would require overcoming the technical, organizational, and regulatory hurdles – essentially, interoperability between IoT systems is critical. Of the total potential economic value the IoT enables, interoperability is required for 40% on average and for nearly 60% in some settings.

Exposing the data via a data hub directly enables maximising the value of vast volumes of multiple types of data – as having the multiple datasets exposed in one place in a uniform manner enables the innovation (see Fig. 5).

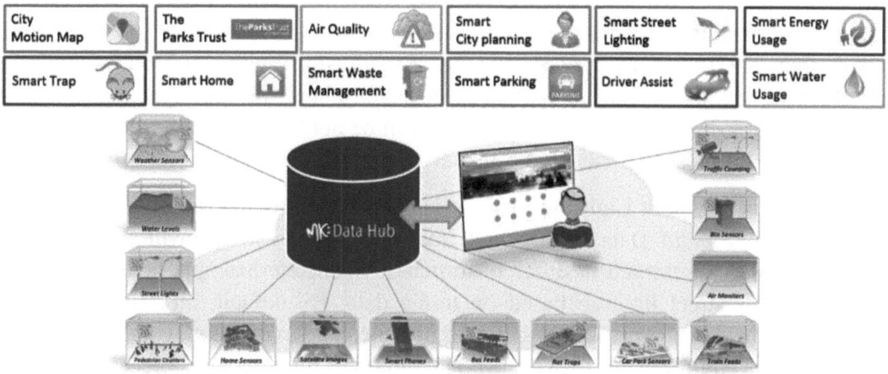

Fig. 5. MK data hub supports innovation across whole ecosystem

Enabling the interoperability between two or more data hubs it is not a simple task. Technical challenges include differences between data hub architecture and endpoint design, authorisation and authentication methods, ways of handling terms and conditions for the data source use, service level agreements and enforcement across the different data hubs. As a first step, interoperability via using Hypercat endpoints is possible. For example a Hypercat enablement on two different data hubs can enable an application developer to query the datasets available utilising the features of the Hypercat API. Providing similar implementations on the consumer API side, might then allow the developer to quickly create a proof of concept using the data sets from different data hubs.

We exemplify by way of several demonstrations, IoT applications which use data from Hypercat-enabled data hubs:

IoT Enabled Bike Share. Within the CityVerve project, the Transport and Travel Theme is looking into challenges of the traditional bike share schemes; investigate how IoT technologies can enhance - and increase awareness of - bike share schemes; and investigate techniques both to increase the amount of new users and to improve user retention rates [11].

Exemplar bike sharing schemes such as in London and Copenhagen are expensive to install and maintain, making them difficult to replicate in cities with less spending power. As there already exists a selection of niche bike share schemes, covering different areas of Manchester, this use case will look into opportunities between existing schemes in terms of enabling effective collaboration in the data layer. The schemes vary in terms of geography, reason for cycling (e.g. commuting, leisure, delivery), cycle aptitude, cycling experience, and bike types (e.g. pedal and electric bikes, commuter use and cargo bikes). The use case will be collecting and aggregating the data from different operators, using various IoT devices; and implementing several distinct business models using those IoT-enabled bikes. The data will be exposed in machine readable manner on the data hub utilising the Hypercat, with a goal of enabling analytics and application layer. With the assets being tracked, optimisation in the operations layer will be combined with the gamification initiatives to both engage citizens to utilise the scheme more but also to assist in the moving of the bikes from where they are to where they should be. By utilising aggregated views on bike availability, bike usage and contextual information about the bike types, user types, paths, parking location, etc., a richer solution can be built. For the end user, the goal is to provide a seamless experience, no matter which bike fleet they chose to use.

Traffic-Sensitive Smart Lighting. Semantic interoperability in general solves the data heterogeneity problem. Data heterogeneity makes it complex to deal with the same data when that data is presented in different formats. A concrete example here is a system which controls street lighting based on traffic data. By default, all street lights are switched off between (say) 12 am and 5.30 am, thereby saving money and energy and benefitting the environment. However, it is required to light certain streets when traffic is on them [12]. As IoT deployments progress, a multiplicity of traffic sensors will be available; including inductive loops at the roadside, Bluetooth-based vehicle counters and cameras. These will report the required data (specifically, the location and speed of vehicles) in multiple formats. By using a semantic description of the data, the system in question does not require additional coding each time a new sensor type is introduced.

Air Quality Management. Semantic interoperability is essential in enabling late binding so that software programs can discover and process data sources not specified a priori. As simple IoT-related example, consider a smartphone app that reports the air quality in a user's current location. As the user changes location, the app can interrogate one or more semantically described IoT data hubs to identify data feeds with data from air quality sensors in the current locality and mechanisms needed how to access that data.

5 Conclusion

Cities are proving to be good exemplars of Internet of Things Ecosystems – thriving with complexity of data sources and the constraints in exposing or using them. While there are many technical challenges to be faced as the IoT continues to develop, the establishment of effective cooperative relationships between multiple ecosystem stakeholders is a significant contributor to the success of IoT applications. In this paper, we have focused on the role of information at the foundation of IoT ecosystems. A number of different roles within the ecosystem were identified, all of which need to derive clear benefit from participation. Projects such as CityVerve and MK:Smart are exploring some of the practical issues associated with creating an IoT ecosystem. Learnings from MK:Smart provided a tangible illustration of the potential of IoT ecosystems, and to identify good practice which can be replicated elsewhere. CityVerve project, with its deployment of different data hubs, shows us that interoperability is essential if the promise of IoT is to be fulfilled. There are already usable standards and specifications which can be applied to IoT systems. The increasing level of activity in standards development organisations and other bodies should identify and address the gaps specific to the new demands of the Internet of Things.

References

1. Manyika, J., et al.: Unlocking the potential of the Internet of Things. McKinsey Global Institute, June 2015. www.mckinsey.com, http://www.mckinsey.com/business-functions/digital-mckinsey/our-insights/the-internet-of-things-the-value-of-digitizing-the-physical-world
2. Davies, J., Fisher, M.: Creating internet of things ecosystems. ITP **9**(4), December 2015
3. Davies, J.: Creating IoT Ecosystems. IET, IET (2016)
4. Beart, P.: Hypercat 3.00 Specification. BSI. (2016). http://shop.bsigroup.com/forms/PASs/PAS-212-2016-download/. Accessed 11 Apr 2017
5. Hypercat Alliance. Hypercat Alliance Web Site. http://www.hypercat.io. Accessed 11 Apr 2017
6. Overview of the General Data Protection Regulation (GDPR). Information Commissioner's Office. https://ico.org.uk/for-organisations/data-protection-reform/overview-of-the-gdpr/. Accessed 11 Apr 2017
7. MK:Smart. MK:Smart Project Web Site. http://www.mksmart.org/. Accessed 11 Apr 2017
8. CityVerve. CityVerve Project Web Site. http://www.cityverve.org.uk/. Accessed 11 Apr 2017
9. Internet of Things Clusters. Innovate UK - UK's innovation agency. https://connect.innovateuk.org/web/internet-of-things-ecosystem-demonstrator/article-view/-/blogs/the-list-of-8-internet-of-things-clusters. Accessed 11 Apr 2017
10. Tachmazidis, I., et al.: A Hypercat-enabled Semantic Internet of Things, March 2017
11. Greater Manchester 2040 Transport Strategy. TfGM. [Online] TfGM, 01 2017. http://www.tfgm.com/2040/Pages/strategy/assets/2017/1-17-0057-GM-2040-Executive-summary.pdf. Accessed 11 Apr 2017

Advances in ICT Research

Architecting for Adaptive Resource Management in Mobile Augmented Reality Systems: Models, Metrics and Prototype Software Solutions

Mykola Tkachuk[1,2(✉)], Oleksii Vekshyn[1], and Rustam Gamzayev[1]

[1] National Technical University "Kharkiv Polytechnic Institute", Kyrpychova Str., 21, Kharkiv 61002, Ukraine
tka@kpi.kharkov.ua, alexeyvekshin@gmail.com, rustam.gamzayev@gmail.com

[2] Karazin National University, Majdan Svobody 4, Kharkiv 61077, Ukraine

Abstract. A 3-level architecting approach to adaptive resource management in mobile augmented reality systems (MARS) is elaborated, which is based on comprehensive data structuring and analyzing of their specific hard- and software features. At the conceptual modeling level an ontology of adaptive MARS resources is constructed, and at the logical modeling level a generic algorithmic model is proposed, which can be instantiated in the collection of specific methods and metrics. As a physical model the reference software architecture for adaptive resource management in MARS is designed, and this approach is implemented partly as a software prototype. It is tested successfully to solve the task of adaptive image resolution on mobile device, according to changes of computational load that finally enables better video stream quality in MARS.

Keywords: Mobile system · Architecture · Adaptation · Resource management · Augmented reality · Model · Metric · Case-based reasoning

1 Introduction

Nowadays mobile information systems can be recognized as more and more efficient and comfortable communication facilities for different kinds of their users. One of the most complex and dynamically grown kind of these systems are mobile augmented reality systems (MARS) [1]. Such systems require more hardware resources than standard mobile applications: e.g. social network clients, instant messengers etc., and this fact leads to maintenance problems of different mobile devices (MD) such as smartphones or tablets. One of the modern trends in software development is using of construction principles for complex systems, especially cybernetic adaptive control schemes for software components, including appropriate decision making models and quality evaluation metrics [2]. These methods are also useful in case of mobile information systems development, in particular for MARS. Such system development approach requires effective usage of restricted resources in MD, and on the other hand supposes to implement complex real-time computational algorithms.

A. Ginige et al. (Eds.): ICTERI 2016, CCIS 783, pp. 17–35, 2017.
https://doi.org/10.1007/978-3-319-69965-3_2

In this paper we propose an adaptive model-based architecting framework for resource management in MARS that includes a collection of domain-specific algorithmic models and quantitative metrics to formalize and estimate some relevant parameters of MARS functioning. The rest of this paper is organized in the following way: Sect. 2 depicts briefly some modern trends in this problem domain and introduces a possible classification of MARS with respect to adaptation issues. Section 3 provides the common vision of model-based architecting concept for adaptive resource management in MARS, and Sect. 4 presents the MARS domain modeling issues using an ontological approach. In Sect. 5 the generic model for resources management in MARS is proposed that includes the collection of appropriate algorithmic models and quantitative metrics to provide an adaptation mode for several MARS subsystems. Section 6 describes the software prototype for adaptive screen resolution facility in MARS, and discusses the experimental data of its testing. Finally, in Sect. 7 a short outlook on the achieved results is discussed, and future work is presented.

2 Mobile Augmented Reality System Development: Classification and Short Review of Related Work

Augmented reality (AR) is a visualized representation form for a real physical environment that is extended by adding computer-generated data [3]. Currently AR supports several data sources: two-dimensional markers; data received from GPS modules (Global Positioning System) [4], and data from build-in gyroscopes. Additionally, MARS could use some modern technologies like image recognition without any markers and GPS data [5].

With respect to the supported data source types, it is possible to construct the MARS classification presented in Fig. 1.

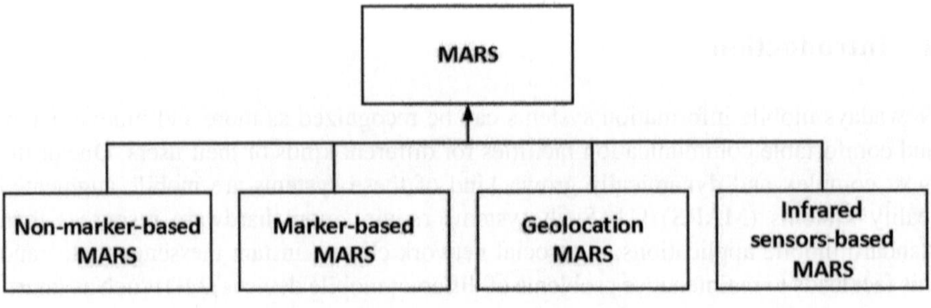

Fig. 1. Classification of MARS implementation technologies

There are four "top-level" types of MARS; each of them has some features and requirements to MD performance. Marker-based MARS operate with special markers which store some data and links to additional information. Non-marker based MARS are more complex type of these systems, they are based on image recognition algorithms, and requires more computational resources to find and recognize free-form objects in an input image. Geo-locational MARS use build-in GPS sensors in order to get information about real environment and augment it with some virtual data. Finally, infrared-sensors based MARS operate with some infrared sensors, which are able to detect objects and move them in a real environment, subsequently this type of MARS is very useful in entertainment and simulators. In scope of this research we are focused below on marker-based MARS, because this class of MARS has the lowest technical complexity, is easy to implement and does not require any additional equipment apart from MD.

In Figs. 2 and 3 the examples of user interface in the demo marker-based MARS are presented [6]. This MARS detects source object in the input video stream recognize it, obtain 3D graphical model of this gear and finally augment source image with this model. Figure 2 demonstrates interface of the marker-based MARS with not augmented image. In current state MARS is ready to analyze source image and search for markers.

Fig. 2. The initial video image [6]

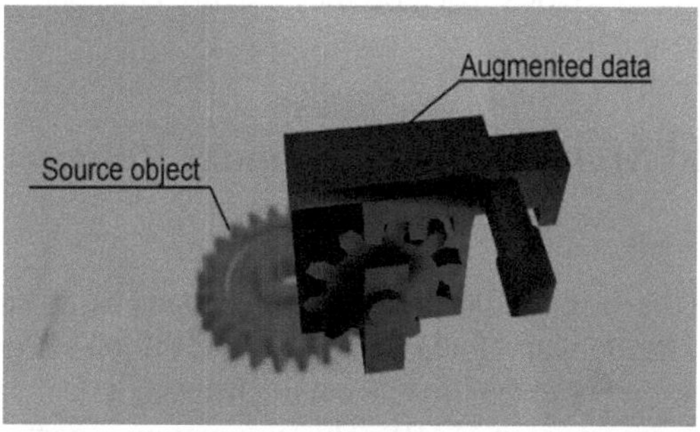

Fig. 3. The example of image with data augmented by marker-based MARS [6]

In Fig. 3 the result image is shown, where MARS finds a marker in the source image (in this case a gear), obtains related data and augments source image with these ones.

In terms of software development there are already some modern software frameworks to develop MARS: Metaio Mobile SDK (Software Development Kit) [7], D'Fusion Mobile [8] and Qualcomm [9].

It has to be noted that last few years a software adaptation became one of the common trends in software engineering, and especially in mobile application development. There are several approaches to adaptation in mobile systems, and some of them were implemented in such projects as [10, 11]: Q-CAD (QoS and Context Aware Discovery), MADAM (Mobility and Adaptation Enabling Middleware) and IST-MUSIC. Recent development is already focused on cloud-based MARS, e.g. CloudRidAR (A Cloud-based Architecture for Mobile Augmented Reality) [12], Elastic Application Model [13], and on common problems of resource management in mobile cloud computing (see e.g. in [14]).

Q-CAD is a resource discovery framework that enables mobile applications to discover and to select resources best satisfied the user's needs. MADAM and ITS-MUSIC frameworks provide a model-driven development approach enabling to assemble applications through a recursive composition process [11]. CloudRidAR project [12] is a cloud-based framework to MARS development which provides development facilities to construct MARS using all advantages of could computing and code offloading, but on the other hand this framework forces developer to use more complex design solutions. Elastic Application Model [13] is based on code offloading, but flexible application architecture and models are built only on the server side called "weblets", and MD hosts only simple client application, which is connected to these components.

More general (not only focused on MARS domain) approaches to resource management problems in mobile cloud computing take into account cloud structure, code offloading and energy efficiency [14]. Besides that, they utilize some economic- and mathematical models: game theory, auction procedures, and optimization methods (e.g., Nash equilibrium concept using in [15], etc.) with corresponding algorithms to calculate the target parameters for adaptive solutions.

To sum up this short review of related work we can conclude that the most part of existing approaches do not provide any complete model-based framework to adaptive resource management in MARS, because they usually consider only some particular aspects of this problem.

3 Model-Based Architecting for Adaptive Resource Management in MARS

Taking into account the results of the provided analysis and based on the under-standing of modern trends in the domain of adaptive MARS-development (see Sect. 2), we can conclude that it is necessary to elaborate a comprehensive model-based framework for adaptive resource management in MARS. This assumption is completely corresponded with such well-proved and recognized approaches in modern software development as model-driven development (MDD) and model-driven architecture (MDA) [16]. Last time these issues are already discussed intensively in a lot of publications about resource management in distributed real-time systems (see e.g. in [17], in SOA- and cloud-centered applications [18, 19], but there is a lack on such work in the domain of MARS development. That is why we propose to provide such model-based architecting for adaptive resource management in MARS framework in the following way:

- to elaborate a domain model to specify all hard and software resources to be analyzed and considered in any adaptive procedure in MARS, and this model represents a conceptual level in the proposed framework;

- to propose some algorithmic-based approaches to manage these resources in adaptive mode, and this vision about the resources in MARS can be considered as a logical modeling level in our framework;
- to develop a collection of reference software architectures to implement a logical model of MARS with appropriate components and interfaces, and such architecting can be represented as physical modeling level in this framework.

The common scheme of these modeling levels is shown in Fig. 4.

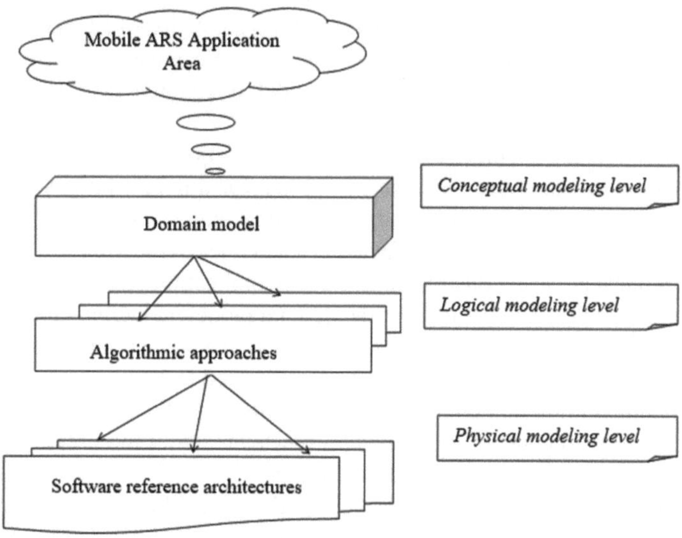

Fig. 4. 3-level modeling framework for adaptive resource management in MARS

According to this model-based view about resource management in MARS for the one and the same domain model a lot of different algorithmic approaches can be elaborated, and for any such approach some different reference software architectures might be implemented. It has to be mentioned that the proposed architecting approach to resource management in MARS completely corresponds, e.g. to well-known and fully recognized 3-level modeling scheme in traditional database development [20].

Basing on the elaborated 3-level modeling scheme (see Fig. 4) and taking into account some specific features of resource management in MARS mentioned in Sect. 2, we propose in this research to consider the following tasks to be solved within our architecting approach to MARS resource adaptation, namely:

1. to provide an adaptive screen resolution (ASR) on mobile devices;
2. to support an adaptive calculation balancing (ACB) between client and server components;
3. to manage an adaptive memory cache (AMC) in run-time mode of MARS.

To solve this list of tasks (1)–(3) it is necessary to elaborate a collection of knowledge-based models, algorithms, numerical metrics, and software solutions that support an adaptive mechanism for resource management in MARS. They are presented in more details in the following sections.

4 Ontological Domain Modeling for Adaptive MARS

Taking into account the 3-level modeling framework for adaptive resource management in MARS proposed in previous Sect. 3, it is needed to elaborate a domain model for this purpose. This model should represent all relevant hard and software capabilities of MD included in any MARS, which in turn can be used as a collection of adaptable parameters for an appropriate algorithmic modeling approach. We have decided to utilize an ontology-based approach to form the domain model for adaptive resource management in MARS.

Ontology models are widely used to represent relationships between concepts in some application domain, and they can be applied for different purposes in software engineering, e.g.: for information sharing between human actors and machines in Semantic Web applications [21]; in software product line engineering [22], and recently ontological models are also constructed in MARS applications for different purposes.

For example, in [23] authors present the domain model including 4 local ontologies like User, Service, Environment and Device to describe general relationships that occur during MARS development. In [24] the ontology is used to build an information basis of educational institution that could be used for rapid MARS development. In [25] the ontologies are used to connect together the knowledge about the users, environment, and business aims in the given application domain.

Most of ontological models mentioned above describe different MARS components taking into account some static resource allocation, and they do not represent system features needed for adaptive resource management. To close this gap in our approach the new MARS domain model was created using the OWL notation, and the OWLGrEd tool is used for this purpose [26], which provides the UML-like graphical editor. In Fig. 5 the proposed ontological domain model for MARS is presented.

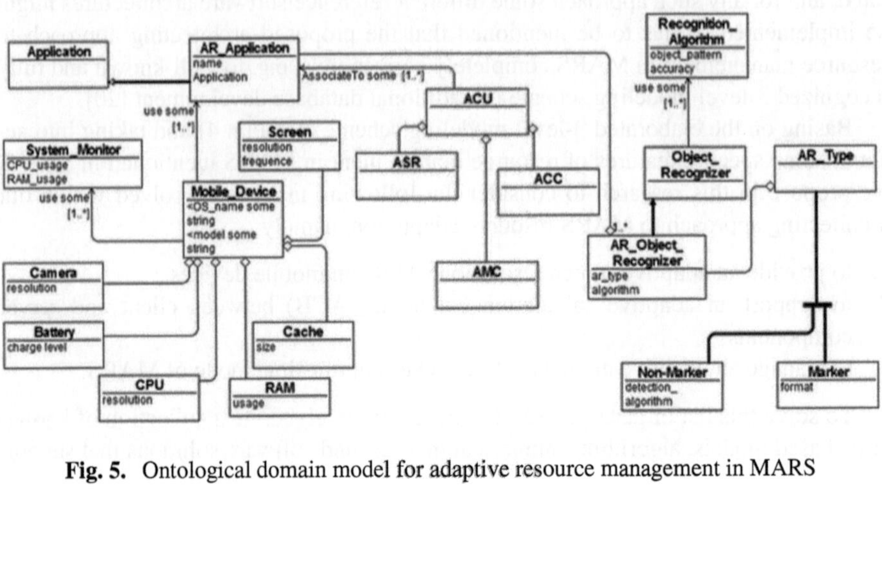

Fig. 5. Ontological domain model for adaptive resource management in MARS

There are the following main entities included in this domain model (see Fig. 5):

1. Augmented Reality Application is an application that uses a technology to analyze elements in the real physical environment and to extend them with additional virtual visual objects or with additional data.
2. Adaptive Control Unit (ACU) is a control facility that is responsible for adaptive resource management in ARS, where the following tasks have to performed: adaptive screen resolution (ASR), adaptive computation balancing (ACB), and adaptive memory cache (AMC).
3. Mobile Device is a small device (iPhone, PDA, notebook, tablet etc.), with restricted amount of system resources, and such a device is managed by an appropriate System Monitor.
4. For resource adaptation of Mobile Device, the following its sub-components (modeling parameters) have to be used: Screen (screen resolution); Cache (cache memory size); RAM (memory size); CPU (computational loading); Battery (charging level); Camera (screen resolution).
5. AR Object Recognizer is a system component to analyze and to extend extracted physical objects with additional or virtual information.

Below this domain model is used consequentially to elaborate all modeling and algorithmic issues for adaptive resource management in MARS.

5 Algorithmic Models and Metrics for Adaptive Resource Management in MARS

According to the proposed adaptive model-based framework (see Sect. 3) at its logical level a collection of algorithmic approaches to adaptive resource management has be constructed with respect to specific hard and software characteristics of mobile devices (MD) operating in a given MARS.

5.1 Generic Algorithmic Model for Adaptive Resource Management

In order to formalize the proposed framework to find adaptive solutions for resource management we can use an algorithmic modeling approach [27, 28], and the appropriate algorithmic model (AM) in a generic way can be defined as the following tuple

$$AM = \langle Workflow\,(Methods),\ InfoBase,\ Metrics \rangle, \qquad (1)$$

where *Workflow(Methods)* are some algorithms which implement the given *(Methods)*, *Infobase* is an information base to be used to perform these methods, and *Metrics* is a collection of metrics to quality assessment of adaptation process in MARS. The choice of adaptation methods in (1) depend on the specific features of MARS resources that should be managed, and one of such possible solutions will be considered in the next subsections. The metrics in (1) also have to be defined taking into account the appropriate hard- and software properties of MD which are used in a target MARS (see domain model shown in Fig. 5).

5.2 Adaptive Screen Resolution Management with Case-Based Reasoning

With respect to the generic algorithmic model defined with formula (1) we propose to elaborate the approach to ASR management with case-based reasoning (CBR) as an adaptation technique.

Taking into account a complex and weak-formalized character of MARS functioning, namely: parallel and multi-threaded computational processes, turbulence overloading on MD, frequent changes on number of users etc., it is reasonable to use so-called soft computing methods [29]: neuronal net technologies, fuzzy-logic methods, generic algorithms, CBR and some others. In particular, exactly CBR-methods can be considered as an effective way to develop decision-making procedures for management of complex software systems (see e.g. in [30–32]). According to this statement the collection of (Methods) in formula (1) can be specified as follows

$$Methods = (NNM, kNNM, kwNNM), \quad (2)$$

where NNM is a Nearest Neighbor Method, kNNM is a k-Nearest Neighbors Method, and kwNNM is k-weighted Nearest Neighbors Method [30].

The main idea of all CBR-methods is that any new problem occurred in some application domain can be resolved using already existing solution for the similar situation (called precedent or case). The CBR-methods differ from each other in a search algorithm to find an appropriate precedent in the given database. For this purpose, it is also important to elaborate an adequate description for the precedents representation, which reflects all relevant issues of MARS functionality.

According to formula (1), *InfoBase* is an information base to apply the CBR-methods defined in (2). It includes a set of precedents, and any such precedent can be defined in the following way

$$c = \left(\vec{p}, \vec{s}\right), \quad (3)$$

where $\vec{p}$ is a vector of parameters to characterize a given problem situation, and $\vec{s}$ is a vector of parameters to represent an appropriate solution for this problem.

Taking into account the hard and software features of MD which are included in MARS (see the domain model in Fig. 5), vector $\vec{p}$ can be given as:

$$\vec{p} = (CPU, RAM, BAT, RES, FPS), \quad (4)$$

where CPU is a current level of a processor loading (in %), RAM is a current level of RAM usage, BAT is a current level of battery charging; RES is a number of possible screen resolution modes in MD, and FPS is a measure of a screen refresh rate.

Further, a vector $\vec{s}$ in formula (2) can be represented as the tuple

$$\vec{s} = (Width, Hight), \quad (5)$$

where *Width* and *Height* are respectively a width and a height of a video frame size on MD.

In [28] is mentioned that a performance of a MARS client application depends on its screen resolution, and accordingly to this reason a number of frame per second (FPS) can be used as one of the metrics from their set *(Metrics)* defined in formula (1). This factor also depends on some parameters: on MD processor performance, on size of its RAM, and on screen resolution of its video camera.

Therefore, a collection of metrics *(Metrics)* in (1) has the following definition

$$Metrics = (T, R), \tag{6}$$

where T is a number of FPS, and R is a total MD's productivity.

A value of metric T can be calculated using the standard function Count(), namely:

$$T = Count(FPS), \tag{7}$$

A value of metric R (named below as a load index) defines a complex characteristic of MD productivity, which is a dimensionless parameter and it can be defined as following

$$R = w_c \frac{CPU_{cur}}{CPU_{total}} + w_r \frac{RAM_{cur}}{RAM_{total}} + w_b \frac{BAT_{cur}}{BAT_{total}}, \tag{8}$$

where CPU_{cur} is a current MD processor loading ratio (in %); RAM_{cur} is a current RAM utilization (in Kb); $BATcur$ is a current battery utilization (in Ah) CPU_{total}, RAM_{total}, BAT_{total} are respectively the nominal values of these parameters; w_c, w_r, w_b are some weighting coefficients for these parameters, and the following condition must be fulfilled $w_c + w_r + w_b = 1$. In this work based on some empirical reasons, we have defined the following value ranges for the index R: it is critical if $R \geq 0.95$; it is high if $0.6 \leq R < 0.95$; it is normal if $0.25 \leq R < 0.6$; and it is low if $0 \leq R \leq 0.25$.

The metrics defined in formula (6)–(8) allow us to estimate the computational resources of an appropriate MD that is used in a target MARS with respect to our final goal: to provide adaptive recourses management in this ARS.

5.3 Complexity Metrics for Adaptive Load Balancing in MARS

As already mentioned above, MARS require more hard- and software resources than other mobile applications, and one of the possible solutions of this problem is an execution of complex business logic on the server side, where computational capabilities are higher than on MARS clients [28]. We propose the approach based on the estimation of computational complexity and analysis of MARS state in run-time. This approach allows to provide an adaptive computation balancing (ACB) in MARS, namely to decide which part of business logic should be executed on the mobile client side, and which one on the server side, depending on CPU (Central Processing Unit) - performance of a MD. In order to provide ACB the collection of metrics to estimate calculation complexity in MARS has been constructed [28].

With respect to the definitions in formula (1), the InfoBase component for ACB can be defined in the following way

$$c = (\bar{a}, \bar{c}),$$

where $\bar{a}$ is precision of calculation; $\bar{c}$ is calculation complexity.

Definition 1. A metric of calculation precision is a parameters vector

$$\bar{a} = (f,\ r,\ s), \tag{9}$$

where f is a number of decimal places after comma; r is an image resolution (in pixels); s is a ratio of image compression (possible values are from 0 to 1).

These values have to be taken into account in the procedure of computation complexity estimation. They implicitly describe possible amount of data for business logic processing, and subsequently affect amount of operations and calculation time on MD.

Definition 2. A metric of calculation complexity is a parameters vector

$$\bar{c} = (c_0,\ c_t), \tag{10}$$

where c_0 is an estimated amount of operations; c_t is an estimated calculation time (in seconds).

Definition 3. A coefficient of a mobile device computation load can be estimated with following expression:

$$P = \frac{D_P \cdot 10^6}{\frac{c_0}{c_t}}, \tag{11}$$

where D_p is the estimated MD performance (in MIPS). It is to notice that because D_p is measured in MIPS, this value should be multiplied by 10^6 to transform its value to c_0 measurement.

Thus, the formula (11) defines the coefficient of the MD computation loading as a ratio of device performance to required amounts of operations in seconds. Taking into account this estimation value, we can conclude that a calculation on MD is possible if and only if $P \geq 1$, otherwise it is necessary to execute the appropriate business logic in MARS on its server side. To illustrate the proposed approach, the following example can be used [28]: let CPU-time's value is $D_p = 0.00961$ *MIPS*, the precision of calculations is $\bar{a} = (3, 96000, 0.6)$, and the estimated complexity of algorithm is $\bar{c} = (3241, 0.4)$. In this case, according to the Eq. (3), value of the coefficient of a MD computation loading is $P = 1.186$, and this calculation should be executed on the MD client side. These issues are discussed in more details in [28].

5.4 An Approach to Adaptive Memory Cache Management

In modern complex client-server software applications, the problem of data access time becomes more and more actual, and one of the possible ways to solve this problem is to use some kind of memory cache component [33]. On the other hand, apart from increasing of data access speed, such component stores most recently used data in random access memory (RAM) that leads also to increasing of a memory usage. Thus, we are facing with the task to choose the most effective memory cache size in order to provide an appropriate data access time with respect to effective usage of RAM.

In our approach to adaptive resource management in MARS we propose a generic vision for adaptive memory cache modeling, and Fig. 6 illustrates this model graphically. To analyze this model, the following parameters, have to be taken into account: free memory rate (M), and amount of unique queries to a database influence on cache capacity (Cs).

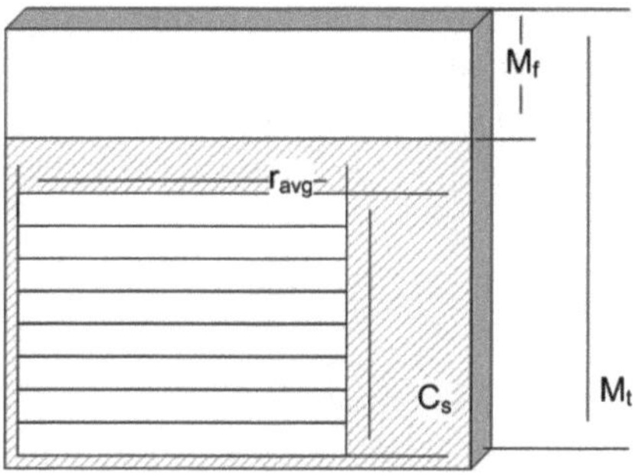

Fig. 6. Graphical interpretation for the approach to Adaptive Memory Cache modeling

We proposed to use CBR method in Adaptive Memory Cache (ACM) as an adaptive management mechanism of, this method allows us to resolve effective cache capacity problem by reusing already existing solution, so it will significantly reduce time for solution finding.

As mentioned before (see Sect. 5.2) a precedent could be defined in the following way:

$$c = (\vec{p}, \vec{s}), \tag{12}$$

where $\vec{p}$ is a vector of parameters to characterize a problem situation, and $\vec{s}$ is a vector of parameters to describe an appropriate solution for problem situation.

In scope of our research we could build $\vec{p}$ and $\vec{s}$ vectors with our cache-specific parameters:

$$\bar{p} = \{ R, M \}, \tag{13}$$

$$\bar{s} = \{ C_s \}, \tag{14}$$

where: *R* is an amount of unique queries (number of items); *M* is a free RAM rate and *Cs* is cache capacity.

But on the other hand, it is necessary to elaborate a way to predict cache capacity with respect to MD state. The most significant parameter for this method is free memory rate M, which is calculated by the following expression:

$$M = \frac{M_f}{M_t}, \tag{15}$$

where M_f – free RAM capacity (MB); M_t – total RAM capacity (MB).

Subsequently, adaptive cache capacity can be calculated with expression below:

$$C_s = R \cdot r_{avg} \cdot M, \tag{16}$$

where *R* is an amount of unique queries (number of items), r_{avg} is an average size of data record stored in cache (in bytes); *M* is a free RAM rate.

Expression 16 gives us a possibility to deal with situation when there is no appropriate precedent in database and we need to create a new one.

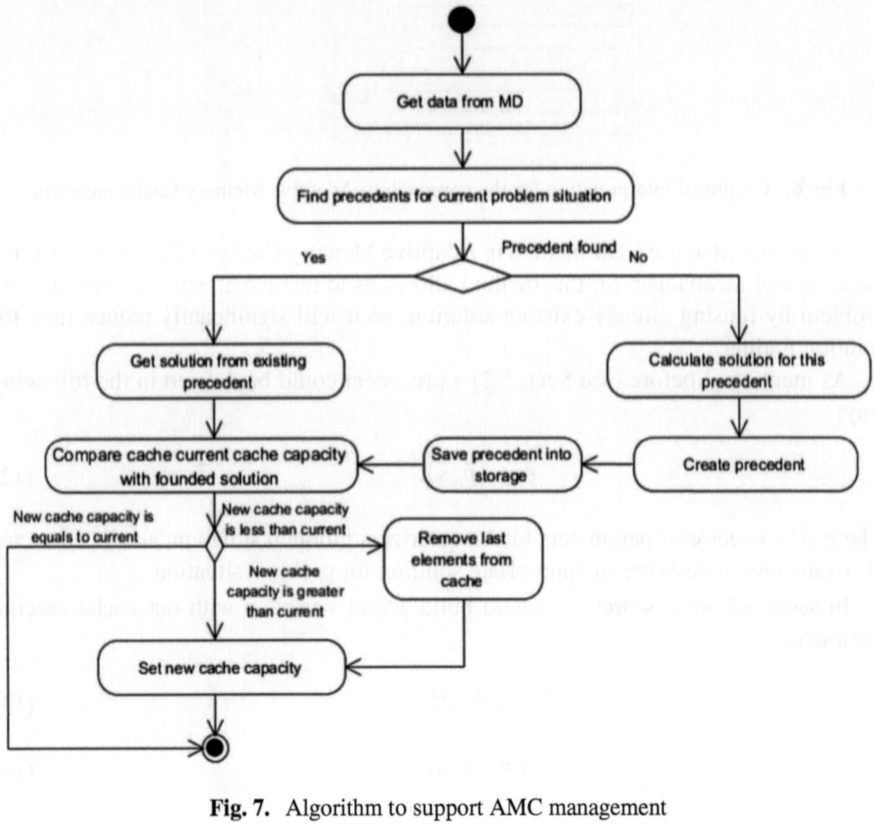

Fig. 7. Algorithm to support AMC management

Basing on CBR method the algorithm for adaptive memory cache management is constructed (see Fig. 7). This algorithm considers the case, when for current problem situation there is no appropriate precedent and a new precedent should be generated, additionally covers actions to perform in a case of increasing or decreasing cache capacity.

The propósed approach for adaptive cache management is quite generic and does not cover all aspects of cache modeling and implementation, but in our future research we will improve this model with more precise components based on modern approaches to cache modeling and construction (see e.g. in [34, 35]).

6 Feasibility Study for Adaptive Screen Resolution Facility: Software Prototype and Experimental Results

6.1 Software Prototype Design and Implementation

In order to depict a feasibility study for the proposed approach the MARS prototype with integrated ACU was developed, but currently with ASR facility only. The core functionality of this prototype is to recognize a marker on some cinema poster in the

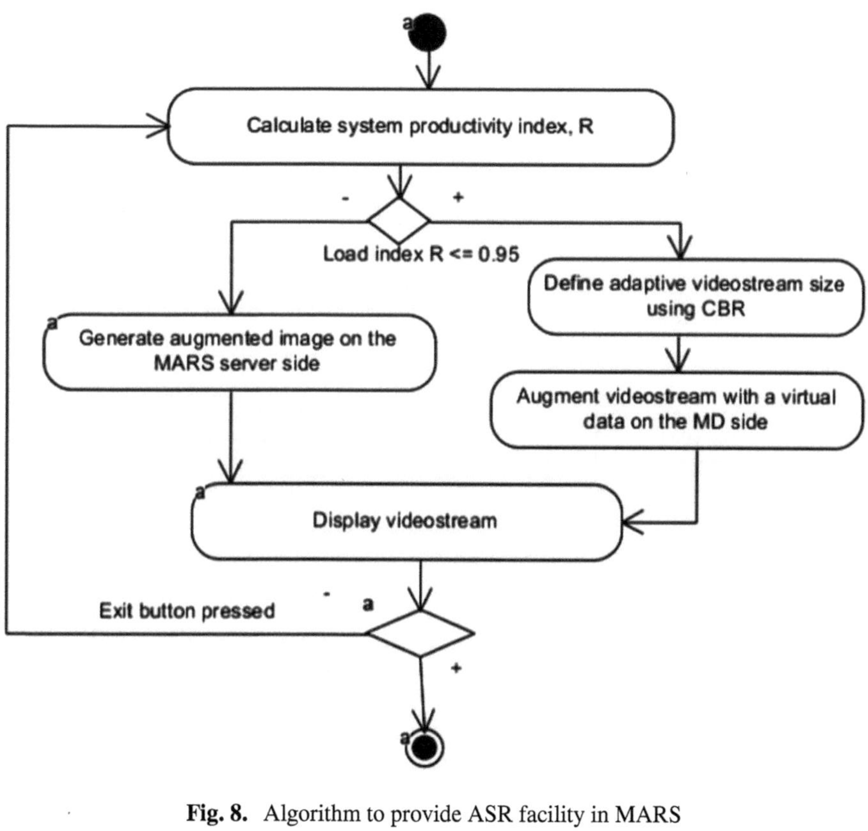

Fig. 8. Algorithm to provide ASR facility in MARS

input video stream, to search some additional data about this cinema in the appropriate database, and to augment recognized cinema with this additional information in real-time mode. Thus, such MARS prototype with integrated ACU is able to analyze environmental parameters and to adapt screen frame size taking into account the MD current state that is given by its load index (see formula (8)). In Fig. 8 the algorithm to provide ASR facility in MARS is presented in form of UML activity diagram [36].

The first step in this algorithm is to calculate the current value of MD's load index. If this value is less than 0.95 ($R \leq 0.95$), it is possible to perform data augmentation on the MD side, so the next steps are to define an adaptive video stream size, using the appropriate CBR method, and to augment the source video stream on MD side, and finally to display the augmented video stream to user. If index $R > 0.95$ it is not possible to augment data on the MD side, and in this case it is necessary to use external resources to augment image from input video stream (e.g. to perform data augmentation on a

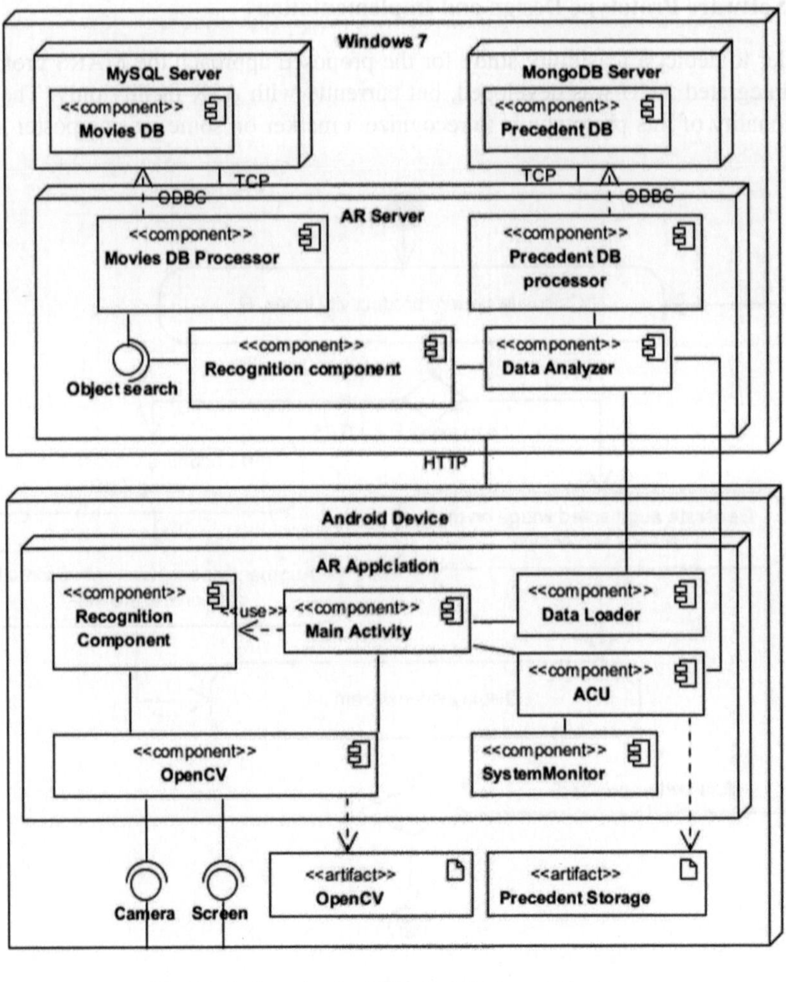

Fig. 9. 3-tier adaptive MARS software architecture

MARS server) and to show the result to a user. The algorithm workflow in both cases can be interrupted by user (in case of Exit button pressed).

To implement the elaborated algorithm, the 3-tier software architecture of MARS was chosen that is presented in Fig. 9 as UML component deployment diagram [36]. This architecture provides some crucial advantages, e.g.: high scalability, data processing security, and lower resource requirements for clients MD.

The client components were implemented with Android platform [37], using embedded Berkeley DB engine [38], and OpenCV library [39]. To develop the server-side components PHP programming language [40], Apache Web-server [41], MySQL [42], and DBMS MongoDB [43] have been chosen.

With respect to the proposed architectural solution, 3 databases have been implemented (see Fig. 9): (1) the local precedents DB (Precedent Storage), this DB is used by the control block ACU; (2) the movies DB (Movies DB), this one stores information about movies, which are processed in MARS prototype; (3) the remote centralized DB (Precedent DB) to store all given precedents, and all local DBs have to be synchronized with this centralized DB. The conceptual data model of the Precedent DB is presented in Fig. 10 as the UML class diagram [36]. This data model takes into account the following entities: Device, Precedent, Param, ListPrecedent, Platform, TypeParameter, to handle all data required by CBR method used in our approach.

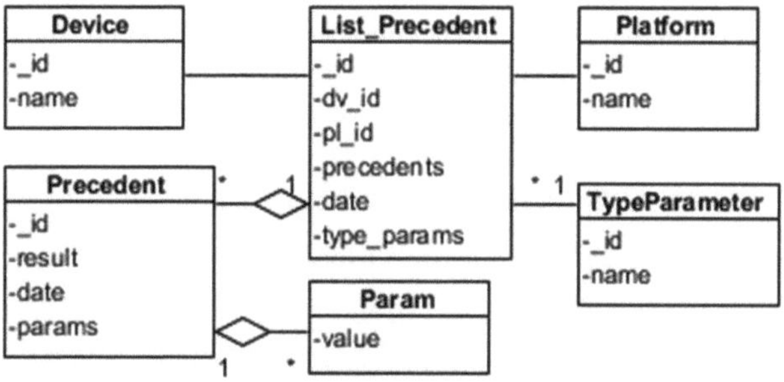

Fig. 10. Conceptual data model of the Precedent DB

To develop this database, we have selected non-relational (No-SQL) database management system MongoDB [43]. This DBMS provides high-speed data processing and stores the data in object-oriented form.

6.2 Test Results and Their Analysis

In order to estimate efficiency of the implemented MARS prototype the experiments have been performed using the following scheme: (1) select mobile devices for experiments; (2) generate precedents DB; (3) run MARS prototype with different operating modes for ACU (with enabled and disabled ACU).

To test MARS prototype two types of MD were selected, the detailed characteristics of these devices are presented in Table 1.

Table 1. Technical characteristics of testing MD

Device name	Processor	RAM	Maximal resolution
Nexus 7	Qualcomm APQ8064 (1.5 GHz)	2 GB RAM	1920 × 1080
Fly IQ4416	MT6572 (1.3 GHz)	512 MB RAM	800 × 600

In Fig. 11 the example of tuples, which describes particular precedents included in the precedents DB, is presented

№	Result	CPU	RAM	BATTERY	RESOLUTION	FPS
1	640x480	0.7	0.7	0.4	800x600	13
2	800x600	0.2	0.5	0.9	640x480	17
3	1024x768	0.3	0.8	0.5	1920x1080	2

Fig. 11. Example of precedents tuple in precedents DB

Two series of experiments have been performed with these mobile devices. The first experiment has been provided with disabled ACU (i.e. without adaptation mode), and the second one with enabled ACU. During these experiments, the image resolution of MD screen in case of different values of productivity index (see Eq. 8) had been measured. In these experiments we took into account 2 intervals from normal and high ranges of the load index: $0.4 \leq R < 0.6$ and $0.6 < R \leq 0.8$. The results of these experiments are presented in Tables 2 and 3 respectively.

Table 2. Experimental results in case of disabled ACU

Mobile device	ëResolution	T for $0.4 \leq R < 0.6$	T for $0.6 < R \leq 0.8$
Nexus 7	640 × 480	15	13
Nexus 7	800 × 600	14	10
Fly IQ4416	800 × 600	10	9
Fly IQ4416	480 × 320	15	14

Table 3. Experimental results in case of enabled ACU

Mobile device	Resolution	T $0.4 \leq R < 0.6$	Resolution	T $0.6 < R \leq 0.8$
Nexus 7	1024 × 768	11	800 × 600	13
Nexus 7	800 × 600	14	640 × 480	16
Fly IQ4416	640 × 480	13	640 × 480	14
Fly IQ4416	800 × 600	9	640 × 480	13

The data from Table 2 show that in case of the fixed image resolution on MD, and if the value of load index R is increased from the values range $0.4 \leq R < 0.6$ to the range $0.6 < R \leq 0.8$, then the T metric (see Eq. (7)) value is decreased, namely, these values are placed in interval {9, 14}; this interval covers values of T metric for both devices: Nexus 7 and Fly IQ4416. In other words, the maximum value's difference T is about

35.7% apart from difference 33,3% for case of {10, 15} and $0.4 \leq R < 0.6$. The reason of such trend in this experiment is the disabled mode of ACU.

Table 3 represents experiment results with the enabled ACU. ACU component monitors computational load on a mobile device and in case of its increasing correct image resolution; such correction leads to stabilization of T metric: this metric changes only in 18.75%. Additionally, in this experiment value of the maximal difference for T in case of $0.4 \leq R < 0.6$ is 37.5%. That is why we can make conclusion that the proposed adaptive resource management approach enables better video stream quality for mobile device in MARS.

7 Conclusions and Future Work

In this paper we have presented the model-based architecting approach to adaptive resource management in mobile augmented reality systems (MARS), which is based on the 3-level data modeling and analyzing of their specific hard- and software features. At the conceptual modeling level, the appropriate ontology-based domain mod-el of all MARS resources was elaborated, at the logical modeling level a collection of algorithmic models and quantitative metrics are proposed, and as a physical model to provide an adaptive resource management in MARS the 3-level reference software architecture is designed and implemented. This approach was successfully applied in order to solve the task to adaptive management of screen image resolution on mobile device according to changes of its computational loading that finally enabled better video stream quality in MARS.

In future we are going to extend a collection of decision search methods in order to improve an adaptation process and compare its accuracy with CBR implementation. It also is supposed to develop a more sophisticated domain model for adaptive MARS with wider amount of system features that should enable a more configure options in the proposed approach. Besides that, using this ontological model we could apply some domain specific modeling methods and appropriate CASE - tools in order to improve finally such important software quality attribute as code reusability (see e.g. in [44]). From the technological point of view, our next step to be done is to apply our architecting for adaptive resource management in cloud-based mobile augmented reality systems.

References

1. Lopez, H., Navarro, A., Relano, J.: An analysis of augmented reality systems. In: Proceedings of the 2010 Fifth International Multi-conference on Computing in the Global Information Technology (ICCGI 2010), pp. 245–250 (2010)
2. Joao, W.C., et al.: Software cybernetics. In: Encyclopedia of Computer Science and Engineering. Wiley (2008)
3. Furth, B.: Handbook of Augmented Reality. Springer, New York (2011)
4. Official U.S. Government information about the Global Positioning System (GPS) and related topics. http://www.gps.gov
5. Official site of CraftAR Service: The Ultimate AR Toolbox. https://catchoom.com/product/craftar/augmented-reality-and-image-recognition/

6. Tkachuk, M., Vekshin, V., Gamzayev, R.: A model-based framework for adaptive resource management in mobile augmented reality system. In: Proceedings of the ICTERI-2016: 12th International Conference on ICT in Education, Research and Industrial Applications: Integration, Harmonization and Knowledge Transfer, Kyiv, Ukraine, 21–24 June, vol. 1614, pp. 41–56. CEUR-WS.org (2016)
7. Official Web-site of Metaio Mobile SDK. http://www.metaio.com/software/mobile-sdk/
8. Official Web-site of D'Fusion Mobile project. http://www.t-immersion.com/products/dfusion-suite/dfusion-mobile
9. Official Web-site of Qualcomm AR SDK. http://www.qualcomm.com/solutions/augmented-reality
10. Kell, S.: A survey of practical software adaptation techniques. J. Univ. Comput. Sci. **14**, 2110–2157 (2008)
11. Kakousis, K., Paspallis, N., Papadopoulos, G.A.: A survey of software adaptation in mobile and ubiquitous computing. J. Enterp. Inform. Syst. **4**, 355–389 (2010)
12. Huang, Z., Li, W., Hui, P., Peylo, C.: CloudRidAR: a cloud-based architecture for mobile augmented reality. In: Proceedings of MARS 2014, pp. 29–34 (2014)
13. Zhang, X., et al.: Towards an elastic application model for augmenting the computing capabilities of mobile devices with cloud computing. J. Mobile Networks Appl. **16**(3), 270–284 (2012)
14. Ionescu, A.: Resource management in mobile cloud computing. Informatica Economica **19**, 55–66 (2015)
15. Alyfantis, G.: Resource management in mobile communication systems and distributed computer systems. Ph.D. thesis, Department of Informatrics and Telecommunication, University of Athens (2012)
16. Sommervile, I.: Software Engineering. Addison Wesley, Boston (2011)
17. Rafiliu, S.: Stability of adaptive distributed real-time systems with dynamic resource management. Ph.D. thesis, Department of Computer and Information Science, Linkoeping Univeristy, Sweden (2013)
18. Ghanbari, H.: Model-based dynamic resource management for service oriented clouds. Ph.D. thesis, York University Toronto, Ontario (2014)
19. Sun, Y., White, J., Eade, S.: A model-based system to automate cloud resource allocation and optimization. In: Dingel, J., Schulte, W., Ramos, I., Abrahão, S., Insfran, E. (eds.) MODELS 2014. LNCS, vol. 8767, pp. 18–34. Springer, Cham (2014). doi:10.1007/978-3-319-11653-2_2
20. Batini, C., Ceri, S., Navathe, S.: Conceptual Database Design: An Entity-Relationship Approach. Benjamin Publishing Company, Redwood City (1992)
21. Fensel, D., Kerrigan, M., Zaremba, M.: Implementing Semantic Web Services: The SESA Framework. Springer, Heidelberg (2008)
22. Tenório, T., Dermeval, D., Bittencourt, I.: On the use of ontology for dynamic reconfiguring software product line products. In: ICSEA 2014, Proceedings of the 9th International Conference on Software Engineering Advances, pp. 545–550 (2014)
23. Hervas, R., et al.: Achieving adaptive augmented reality through ontological context-awareness applied to AAL scenarios. J. Univ. Comput. Sci. **19**(9), 1334–1349 (2013)
24. Chuan-Jun, S., Tzu-Ning, Y., Yu-Ming, Y.: Ontology-based mobile augmented reality for personalized U-Campus. In: Proceedings APIEMS 2012, pp. 2037–2046 (2012)
25. Hatala, M., Wakkary, R.: Ontology-based user modeling in an augmented audio reality system for museums. J. User Model. User-Adap. Inter. **5**(3–4), 339–380 (2005)
26. Bārzdiņš, J., et al.: OWLGrEd: a UML Style Graphical Notation and Editor for OWL 2. In: Proceedings of 7th International Workshop "OWL: Experience and Directions" (2010)

27. Ramesh, K., Karunanidhi, P.: Literature survey on algorithmic and non-algorithmic models for software development effort estimation. Int. J. Eng. Comput. Sci. **2**(3), 623–632 (2013)
28. Vekshyn, O., Tkachuk, M.: Algorithmic software adaptation approach in mobile augmented reality systems. In: Proceedings of 7-th International Conference on Software Engineering Advances, Lisbon, Portugal, pp. 40–43 (2012)
29. Aliev, A.: Soft Computing and its Applications. World Scientific (2001)
30. Maiden, N., Sutcliffe, A.: Case-based reasoning in software engineering. In: IEE Colloquium on Case-Based Reasoning, Digest No. 036, pp. 1–3 (1993)
31. Limthanmaphon, B., Zhang, Y.: Web service composition with case-based reasoning. In: Proceedings of 14th Australian Database Conference (ADC 2003), vol. 17 (2004)
32. Tkachuk, M., Polkovnikov, S., Bronin, S.: Adaptive control framework for software components: case-based reasoning approach. In: Proceedings of 6th International Workshop on Software Cybernetics (IWSC–2009), Seattle, USA, pp. 47–56 (2009)
33. Hankins, R., Patel, J.: Data morphing: an adaptive, cache- conscious storage technique. In: Proceeding of the 29th VLDB Conference, Berlin, Germany, pp. 417–428 (2003)
34. Press, A., et al.: Caching algorithms and rational models of memory. In: Proceedings of the 36th Annual Conference of the Cognitive Science Society, Quebec City, Canada (2014)
35. Bender, M., et al.: Cache-adaptive analysis. In: Proceeding of the 28th ACM Symposium on Parallelism in Algorithms and Architectures, Pacific Grove, California, USA, pp. 135–144 (2016)
36. Official Web-site of OMG Consortium. http://www.uml.org/
37. Official Web-site of Android platform. https://www.android.com/
38. Official Web-site of Oracle Berkeley DB. http://www.oracle.com/technetwork/database/database-technologies/berkeleydb
39. Official Web-site of OpenCV project. http://www.opencv.org
40. Official Web-site of PHP language. http://www.php.net
41. Official Web-site of Apache project. http://www.apache.org
42. Official Web-site of MySQL project. http://www.mysql.com
43. Official Web-site of MongoDB. http://www.mongodb.org
44. Tkachuk, M., et al.: An integrated approach to evaluation of domain modeling methods and tools for improvement of code reusability in software development. In: Heinrich, C., Mayr, Pinzger, V. (Hrsg.) INFORMATIK 2016, Lecture Notes in Informatics (LNI), vol. P-259, pp. 143–156. Kollen Druck+Verlag GmbH, Bonn (2016)

Intelligent Cyber Defense System Using Artificial Neural Network and Immune System Techniques

Myroslav Komar[1(✉)], Anatoliy Sachenko[1,2], Sergei Bezobrazov[3], and Vladimir Golovko[3]

[1] Research Institute for Intelligent Computer Systems, Ternopil National Economic University, Ternopil, Ukraine
{mko,as}@tneu.edu.ua

[2] Silesian University of Technology, Zabrze, Poland

[3] Brest State Technical University, Brest, Belarus
bescase@gmail.com, gva@bstu.by

Abstract. Over the past few decades, the application of Artificial Immune Systems (AIS) and Artificial Neural Networks (ANN) has been growing rapidly in different domains. We sincerely believe that integration of these both techniques can allow constructing the Intelligent Cyber Defense System. In this paper an original method for detecting the network attacks and malicious code is described. The method is based on main principles of AIS where immune detectors have an ANN's structure. The main goal of proposed approach is to detect previously unknown (novel) cyber-attack (malicious code, intrusion detection, etc.). The proposed Intelligent Cyber Defense System can improve the reliability of intrusion detection in computer systems and, as a result, it may reduce financial losses of companies from cyber attacks.

Keywords: Artificial neural networks · Malicious code detection · Intrusion detection · Intelligent system · Cyber attacks · Cyber defense · Financial losses

1 Introduction

The up-to-date computer system cannot be imagined without safety equipment. The work in the Internet is accompanying by high risks to be attacked by network intrusions and malicious code. As a result, the cybercrime continues to do more financial damage to companies: a company's costs for preventing cybercrimes are estimated approximately $ 15 million per year. The CIS Company [1] reported that during last years every second organization was attacked. The costs of each company may vary from $ 1.9 million to $ 65 million per year. In absolute terms, the damage from cyber attacks increased by 82% in the last six years [2]. In 2011, the direct annual global losses from cybercrime were estimated $ 114 billion. Taking into account the financial losses of companies from cyber attacks and the costs of downtime and recovery, the cybercriminal activity is worth of $ 388 billion per year to the world economy [3].

A. Ginige et al. (Eds.): ICTERI 2016, CCIS 783, pp. 36–55, 2017.
https://doi.org/10.1007/978-3-319-69965-3_3

All cyber-attacks can be divided into two large groups: network attacks and malicious code attacks.

According to the opinion of Forbes experts [4], one of the high-profile crimes in the field of information security was the Anonymous attack on the MasterCard, Visa and Paypal payment systems in late 2010. Damage from this attack was worth of $ 5.5 million. The other high-profile cybercrime was an attack on Citibank in June 2011. Hackers had stolen $ 2.7 million from the accounts of 3,400 customers of the bank. Breaking into Sony PlayStation Network took place in April 2011. The total damage to the companies was estimated $ 171 million.

There are a tremendous number of cyber attacks which have a growing trend. In particular, in the third quarter of 2016 «Kaspersky Lab» blocked more than 171 million of malicious attacks on user computers and mobile devices all over the world. At that, web antivirus detected more than 12 million of unique malware (scripts, exploits, executable files, etc.) and antivirus software found more than 116 million of virus attacks [5].

The quality and complexity of malware are constantly enhancing. The striking example of such new cyber threat is Stuxnet [6]. It was discovered in July 2010 but experts are still analyzing the ability and hide functions up till now. Stuxnet was written especially for attacking Supervisory Control and Data Acquisition (SCADA) systems used to control and monitor industrial processes. Stuxnet includes the capability to reprogram the PLCs, to take over control of object and to hide its changes. This malware probably has damaged Iran's nuclear facilities in Natanz and eventually delayed the starting of Iran's Bushehr Nuclear Power Plant. Several security companies claim that Stuxnet is "a working and fearsome prototype of a cyber-weapon that will lead to the creation of a new arms race in the world" [7].

The users of social networks also undergo an often cyber attacks. The main goal of such attacks is the stealing of the confidential information and placement of links on infected web resources. So, for instance, the worm Koobface [8] displays the largest activity, attacking Twitter and sending messages with links to Trojans.

Throughout several last years a hack group Anonymous together with others was actively involved in various operations against law enforcement agencies, banks, governments, security companies and major software vendors. All these led to incidents with security breaches of networks belonging to the United Nations, a security intelligence firm Stratfor, FBI contractor IRC Federal, US Defense contractor ManTech, and the CIA [9].

In August 2011 the malicious code named Duqu [10] was detected. This Trojan is actually a sophisticated attack toolkit, which can be used to breach a system and then systematically siphon information out of it. New modules can be uploaded and run on the fly, without a file system footprint. The highly complex and intricate modular architecture, made Duqu undetectable for years. Some experts are of opinion that "Duqu and Stuxnet represent the state of the art in cyber warfare and hint that we are entering an era of cold cyber war, where superpowers fight each other unconstrained by the limitations of real-world war" [11].

As it states above an urgent problem is to have effective methods defending against cyber-attacks.

2 State-of-the-Arts

Depending on used techniques, experts define four basic types of network attacks: denial of service attacks, user-to-root attacks, remote-to-local attacks and probe attacks, and several subtypes of these attacks [12, 13].

Nowadays many methods for solving the problem of network attacks detection where developed. The essential part of these methods is based on artificial intelligence such as artificial neural networks, methods of fuzzy logic and artificial immune systems.

Artificial neural networks are characterized by the learning ability, and generalizing of learning outcomes, and ability of the noisy process and incomplete data. This is the reason that ANN found the broad implementation for solving the problems in the area of the cyber defense, in particular for intrusion detection [14].

Intrusion Detection Systems (IDS) on the ANN can be divided into four categories [15]. The first category (earlier studies) of IDS is based on Multi-Layer Feed Forward Neural Network (MLFF), such as the Multi-Layer Perceptron (MLP) and Back Propagation (BP). Thus, Cannady [16] applied the MLP to analyze the network traffic for the external attacks identification. In [17] the MLFF was used to detect anomalies caused by user behavior. Authors [18, 19] employed the MLP to detect anomalies, and in [20] the Time Delay Neural Network was used. A comparison of MLFF with other neural networks has made [21, 22], and it's established that MLFF neural networks have a lower efficiency than Kohonen Self-organizing map [23].

The second category of IDS is based on Cerebellar model Articulation Controller (CMAC) neural networks and Elman neural networks. Cannady [24] proposed to apply the CMAC for online learning and adapting to novel attacks. Debar et al. [25] employed a simplified Elman recurrent neural network and recurrent multilayer back-propagation network to predict instructions. Cheng et al. [26] designed a model for detecting anomalies in network traffic based on Elman recurrent neural network as a classifier of network attacks. This approach enabled to improve the results of detecting the novel network attacks.

The third category of IDS is based on unsupervised learning the neural networks for classifying and visualizing the input data to recognize the normal behavior from abnormal one. Most systems in this category use the Kohonen Self-Organizing Map (SOM). Höglund et al. [27] and Ramadas [28] trained the SOM in collected normal data and employed it to detect the abnormal activity of users. Authors [29] grouped 41 parameters of the network connection into three groups and trained the three-layer SOM on dedicated parameters that enabled to reduce essentially the error of first kind.

The fourth category of IDS is based on hybrid neural networks. Jirapummin [30] proposed a hybrid neural network including the SOM for intrusion detection and a network of «resilient propagation» (RPROP) for intrusion classification. Horeis [31] used a combination of SOM and Radial Basis Function (RBF) neural network, and it showed better results than the system based on RBF networks only.

Intrusion detection models on fuzzy logic are using the fuzzy rules or fuzzy classifiers [32]. Dickerson et al. [33] proposed a Fuzzy Intrusion Recognition Engine (FIRE) for the detecting the malicious activity in the network. Data portions are classified using static metrics and enabling to generate fuzzy rules for classification of the input network data. The main disadvantage of this approach is that the rules are created manually, but not automatically. Moreover the process of rules generating is laborious, and it imposes serious constraints on system development.

In the field of Artificial Immune Systems several basic algorithms were proposed: Negative Selection algorithm [34–36] Clonal Selection algorithm [37, 38], Idiotypic Network [39, 40] and Dendritic Cell algorithm [41]. Let's consider the most famous AIS algorithms: Negative Selection and Clonal Selection.

A. Perelson and S. Forrest proposed the Negative Selection algorithm for solving the anomaly detection problems [34–36]. It's based on the process of lymphocytes maturation in the thymus – biological organ that plays the basic role in the human immunity [42, 43]. The negative selection of T-cells in the thymus consists of maturation and selection processes for thymocytes (immature T-cells). During the selection process, Antigen Presenting Cells (APCs) present self-peptide/major histocompatibility complex (MHC) to the T-cells. Those cells - that react strongly (bind with high affinity) with the self-peptide/MHC complexes - are eliminated through a controlled cell death, called apoptosis. As a result, only such T-cells remain and they can recognize foreign antigens that are not self-reactive.

Despite its successful application, the negative selection algorithm has several serious weaknesses [40, 44]. The first, it needs to create a randomly-generated initial detector population. If the dimensionality of the future space increases, then a number of detectors is growing exponentially. The second, the definition of "normal" is not updated as the time progress. And the third, the negative selection algorithm can cause excessive numbers of false alerts.

F. Burnet proposed the Clonal Selection algorithm based on the Clone Selection theory [37, 38]. This theory explains the basic response of the adaptive immune system to an antigenic stimulus during proliferations of B-cells. When receptors (antibodies) of B-cells bind with an antigen, B-cells become activated and differentiate into plasma or memory cells. Generated clones of B-cells are undergoing somatic hyper-mutation, thus introducing diversity into the B-cell population. Plasma cells produce large numbers of antigen-specific antibodies that lead to the removal of the antigen. Memory cells live in the organism for a long time and provide a rapid secondary response upon a subsequent encounter with the same (or similar) antigens. The algorithm of Clonal Selection is usually applied for developing the optimization approaches.

In our opinion, main problems of most AIS applications for data mining and anomaly detection tasks are the complex structure of immune detectors and no representatively matching methods. For example, in [34, 45] the binary structure of detectors is employed. Such structure of detectors requires the use of a contiguous bit matching method (r-contiguous bit [34], r-chunks [46]) that reduces the space and time complexity. Several works [44, 47] outline the unacceptable computational complexity of such methods by reason of the exponential relationship between the size of the data set (to be used) and the number of detectors that it is possible to generate. Also Gonzalez et al. [48] showed that matching rules between two binary strings cannot represent a good generalization of a self-space, and detectors demonstrate the insufficient good coverage of a nonself-space [49].

In comparison with mentioned methods above we propose the immune detectors based on neural network. An artificial neural network is an adaptive system that changes its structure based on external or internal information, and it flows through the network during the learning phase, and it's characterized by learning capability, generalizing ability and self-organization. Implementation of ANN enables to avoid the listed weaknesses above and increases the self-adaptation and self-evolution abilities of detectors in the tasks of data mining and anomaly detection.

In this paper we investigate the ability of immune detectors with neural network architecture to adapt to the changeable software environment and self-evolution in order to detect the novel threat. The adaptability and self-evolution of detectors consists in modification of its structure for increasing the detection rate of novel cyber attacks.

The rest of the paper is organized as follows. Section 3 explains the generalized structure of developed system. Section 4 presents a mechanism of adapting the immune detectors to the new detected cyber attacks. The results of experiments are discussed in Sect. 5. Finally, Sect. 6 contains Conclusion and Future Work.

3 Generalized Architecture of Intelligent Cyber Defense System

The Artificial Immune System for Cyber Defense is the set of "intelligent" immune detectors (the structure of immune detectors, the algorithm of their training and evolution are described in Sect. 4) and rules that describe their behavior. The system consists of modules that perform the control of immune detectors. The immune detectors are going through the different stages during the lifetime. There is creation, training, selection, detection etc. stages. Each stage can be represented as a module of the defense system. Thus, the developed system for computer attacks detection consists of several interacting modules (Fig. 1).

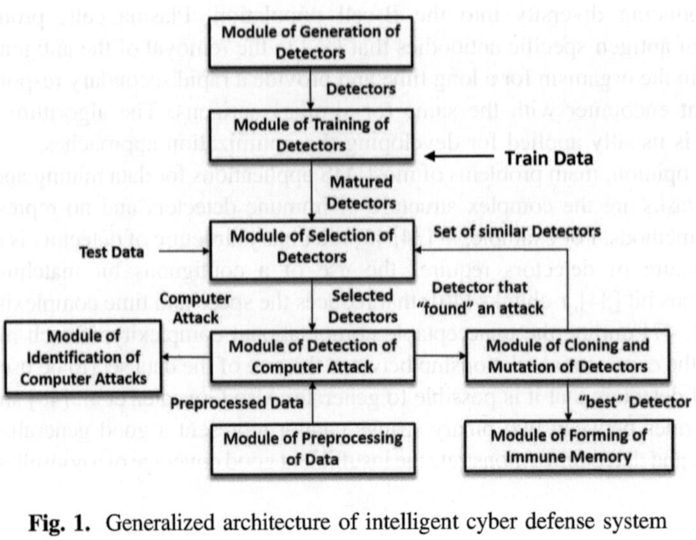

Fig. 1. Generalized architecture of intelligent cyber defense system

The module of generation of detectors produces the set of so-called pre-detectors that go through the several stages before they acquire the ability of correct classifications of objects. Every immune detector has the limited lifetime during which it "lives" in the system. At the end of the lifetime the detector is replaced by another new detector. This mechanism provide the system the continuously inflow of new immune detectors with different structure and different ability that can be more powerful than its precursors.

After creation, during the train stage, the immune detectors acquire the ability of correct classification of different objects and processes in computer environment and to detect the cyber-attacks. The train set contains the pattern from the both classes – legitimate and threat and its structure depends on observed data. It can be data from executable files in a case of malware detection and the parameters of network connections in the case of detecting the network attacks. The process of train data creation for the malicious code detection task is described in the Sect. 5.2.

After training all immune detectors going through the selection stage where detectors pass the checking for correctness to minimize the erroneous work, while legitimate objects (files, processes, connections etc.) is considered as the threat. For this purpose, the preliminary created test sample – which consists of legitimate objects only – is given to detectors. If a i-th detector classifies one of test objects, as an attack, then it is destroyed and replaced by a new detector. If a i-th detector does not generate the erroneous work during the test selection, then it is considered as a correct one and admitted to the analysis of computer environment. As a result the set of immune detectors for the analysis of environment is created, and it can be filled up due to the detectors of immune memory and generating new detectors after the end of their lifetime. The module of selection enables to decrease the false alarm rate and as well as the defense level.

All selected detectors can defense a computer system against cyber attacks. The set of active immune detectors forms a multi-agent system, where each immune detector is an intelligent agent with its own list of tasks. It selects the target of scanning, makes clones. At the functioning stage (or detection stage) all the information – which is getting by the computer – is primarily analyzed by immune detectors. If none of detectors found an anomaly, then data are processed by the operating system and the proper software. In addition, some period of life is given to each detector which analyses the environment during given period. If upon termination of the given time the detector didn't find the anomaly, it is destroyed, and a new detector is created but on its place. If any object is classified by immune detectors as the attack, then such detectors react on this attack. For example, they can block the proper connection, and, as a result, it is not processed by the operating system and software. The user receives the message simultaneously about the attack attempt on the computer system.

If detector found a threat then processes of cloning and mutation are activated. The goal of cloning and mutation module is to produce copies of the immune detector that found the attack. Such "clones" that are similar to the "parent" are very useful for example for defense against the malware family, where each example has the similar malicious code. Such clones are capable to react on the found malware and check all the objects in the computer environment for a short period.

When clones are creating, some changes in their structure are taking a place. As a result, the clones are not exact copies of the parent but with small differences. This process is called 'mutation' (mutation is described in Sect. 4 in details). It enables immune detectors to acquire the new ability, adapt to novel attacks and increase the detection rate. In our case, when detectors base on neural networks, each clone is training on information abstracted from the detected attack. It allows adapting to novel attack with the purpose of increasing the quality detection. Section 5.2 contains details about cloning process.

During the detection and eliminating of attacks, it is expedient to save their parameters and samples for the further detailed analysis. The point is that immune detectors are trained on the limited set of data, which can not include all possible cyber attacks. Therefore, all samples of attacks, which were classified as the novel ones, are saved and added to the trained sample. It enables to increase the authenticity of attacks detection and classification as well as provide the system flexibility, so this process updates the information. Newly generated detectors will be trained already, on new data.

At the end, the best detector is chosen and transformed to the "memory" detector. Memory detectors have the unlimited lifetime and provide a quick reaction on repeated cyber attacks. Thus the set of memory detectors forms the "immune memory", and keeps the information about all met cyber attacks, and provides the high level of reaction on repeated attempts of attacks.

Finally, the module of threats identification is used for classifying the detected threat. The knowledge about the class of the detected threat enables making the correct response.

4 Structure of Immune Detector

Selecting the structure of the immune detector is very important since it influences directly to the detection ability. In our opinion the neural network structure of immune detectors is preferable and it enables to construct more powerful detectors. We designed immune detectors which are based on the feed forward counter propagation neural network [50]. The counter propagation network is guaranteed that it finds the correct weights during the learning process (Fig. 2).

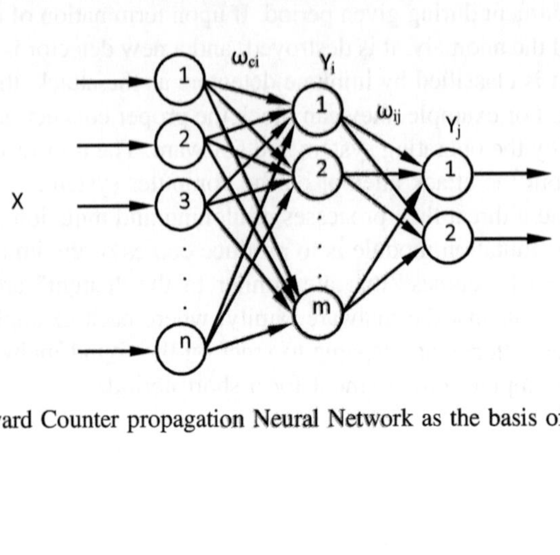

Fig. 2. Feed forward Counter propagation Neural Network as the basis of Immune Detector

The neural network includes the three layers of nodes. Input layer's nodes connect to each node in the hidden layer and receive data from outside (it can be different files, system processes, network traffic and etc.). The number of inputs nodes n defines the size of the window for data inputs to neural network.

The hidden layer consists of m Kohonen neurons and it represents a vector quantization layer [23, 51], which gives the cluster label of the input pattern. The competitive learning rule (winner-takes-all) is used for training the hidden layer. The number of neurons

$$m = p + r \tag{1}$$

where p is the number of the first neurons which corresponding to legitimate files;

r is the number of last neurons, their activity characterizes the class of malicious files.

The ratio of p to r should be multiple of 4 to 1 (for example $p = 8$, $r = 2$).

This ratio is related to the algorithm of forming the learning sample which was received experimentally and showed best results.

The output layer consists of linear units and carries out mapping of clusters into classes. It includes the two nodes: the activity of the first node indicates the legitimate object; the activity of the second node represents a computer attack. Weights between layers and relationships and classes are defined within the learning phase.

During the lifetime the immune detector goes through a several stages: creation, learning, selection, cloning and mutation, notably the immune detector evolves during its "life" [52, 53].

The detectors' life starts from its creation. At this stage the neural networks with random weights are generated.

At the training stage the created neural network immune detectors are subjected to the training process. As a result the ensemble of various detectors is created, and each detector can detect different computer attacks.

Let's consider the N is the data which belongs to the certain type of computer attacks, and the M is the data which belongs to the class of legitimate objects. Then the ensemble of input images can be formed casually for the i detector training:

$$X_i = \begin{bmatrix} X_i^1 \\ X_i^2 \\ \cdots \\ X_i^L \end{bmatrix} = \begin{bmatrix} X_{i1}^1 & X_{i2}^1 & \cdots & X_{in}^1 \\ X_{i1}^2 & X_{i2}^2 & \cdots & X_{in}^2 \\ & & & \cdots \\ X_{i1}^L & X_{i2}^L & \cdots & X_{in}^L \end{bmatrix} \tag{2}$$

Accordingly, we get a plural of reference images

$$e_i = \begin{bmatrix} e_i^1 \\ e_i^2 \\ \cdots \\ e_i^L \end{bmatrix} = \begin{bmatrix} e_{i1}^1 & e_{i2}^1 \\ e_{i1}^2 & e_{i2}^2 \\ & \cdots \\ e_{i1}^L & e_{i2}^L \end{bmatrix} \tag{3}$$

where L is a dimension of the training sample.

Reference output values for the i detector are formed as the following:

$$\begin{aligned} e_{i1}^{k} &= \begin{cases} 1, & \text{if } X_i^k \in N \\ 0, & \text{otherwise} \end{cases} \\ e_{i2}^{k} &= \begin{cases} 1, & \text{if } X_i^k \in M \\ 0, & \text{otherwise} \end{cases} \end{aligned} . \quad (4)$$

First of all a training of neuron network is running to the moment of the total quadratic error minimization:

$$E_i = \frac{1}{2} \sum_{k=1}^{L} \sum_{j=1}^{2} (Y_j^k - e_j^k)^2, \quad (5)$$

where Y_j^k is the j output value of detector for the k input image;
e_j^k is the j standard value for the k reference image.

Then the training is running unless the amount of trained immune detectors becomes equal to the set value of F.

After training the set of detectors scan an environment and classify objects. The proposed structure of immune detectors enables to use the small dataset for training and classify correctly real-world patterns after training.

During the lifetime neural network immune detectors evolve continually. The evolution of immune detectors is an important part the intelligent system running, because it enables to expose new regularities and features of continually appearing novel attacks as well as adapt to it. As a result the system is evolving and improving its defense's abilities. Let's examine the process of adaption.

If the i-th immune detector detects the attack then it activates the alarm, and the cloning and mutation of the given detector are performed. As a result the set of clones is generated and each clone is trained using the attack code (the process of mutation).

The algorithm of evolution consists of following steps:

- Creating a set D of copies (clones) of the detector that found the computer attack.
- Creating the learning sample L from the data of the detected attack.
- Training the clones.
- Calculating the fitness F of clones. If the fitness increases the clone then it is "good". In the opposite case the clone is eliminated.

As a result, the set of detectors-clones D_i are produced, which are aimed to detect the given attack.

The goal of the mutation is to explore a novel computer attack, and find samples of novel attack's techniques, and elaborate robust detectors. Therefore we apply the relearning process of clones. During this process clones adapt to the novel attack and provide an effective defense from the active cyber attack.

The fitness function is used to determine the detection quality. The mean-squared error between input and output vectors (5) for the immune detector can be used as the fitness.

The immune memory is creating in the final step of immune detectors evolution. The immune memory consists of the "best" neural immune detectors which confirmed the highest fitness during the detecting the certain cyber attacks

$$M_k = D_i,\ if\ E_i < E_j. \tag{6}$$

Detector-clones with the minimal value of mean-squared error are transformed into the memory detector with the "unlimited" lifetime.

5 Results and Discussions

5.1 Ability of Artificial Cyber Defense System Detecting the Network Attacks

The aggregate classifier based on neural network immune detectors [52–54] is offered to identify and hierarchically classify computer attacks. Each detector has been trained on a specific type of the attack that allows determining the attack classes and types. The proposed classifier is based on joint application of the Principal Component Analysis (PCA) [55] and on resolving the joint conflicts between neural network immune detectors trained on a specific type of the attack. A technique for dimensionality reduction of the input data space by means of the software implementation of the PCA was considered in [56]. Conducted researches confirmed that one of the main components contains 52.4% of the information; the two main components contain 71.7% of the information; the three main components contain 88.4% of the information, and the 12 first principal components contain 99.2% of information about network connections.

The PCA can be carried out on the basis of the deep neural network (autoencoder) [57–59]. The number of neurons in the input and output layers of the proposed autoencoder (Fig. 3) corresponds to the number of network connection parameters of «KDD Cup'99» base [12] and is equal 41. The number of neurons in the hidden layers is determined empirically like training options.

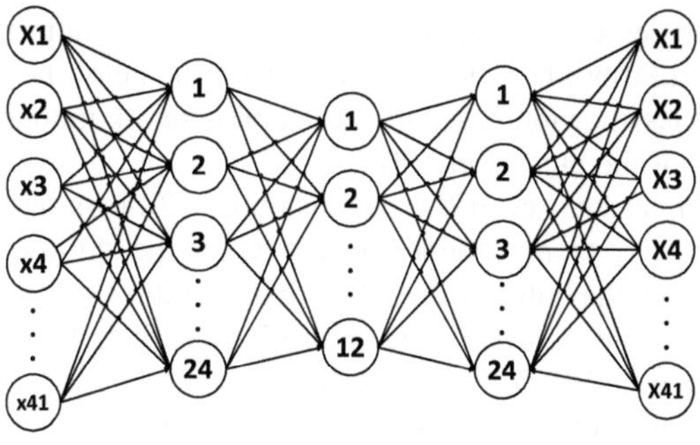

Fig. 3. Autoencoder structure for Principal Component Analysis implementation

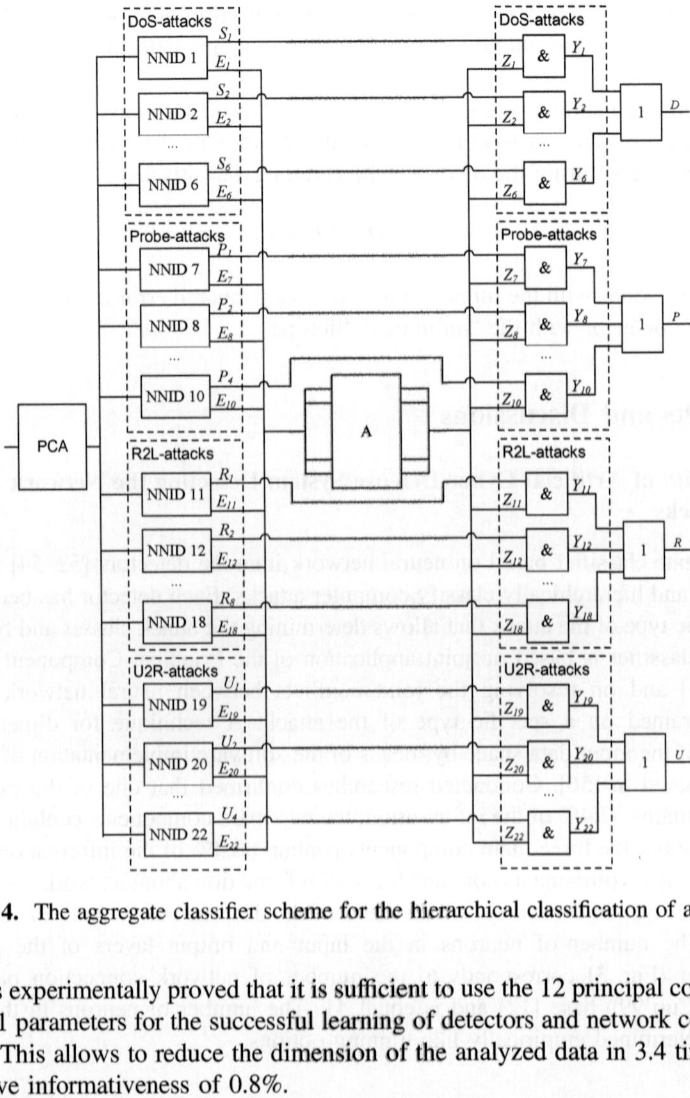

Fig. 4. The aggregate classifier scheme for the hierarchical classification of attacks

It was experimentally proved that it is sufficient to use the 12 principal components, but not 41 parameters for the successful learning of detectors and network connections analysis. This allows to reduce the dimension of the analyzed data in 3.4 times losing the relative informativeness of 0.8%.

Having the 22 types of the defined network attacks combined into four classes (DoS, U2R, R2L and Probe) [12], let's consider the scheme of the aggregate classifier for the hierarchical classification based on 22 neural network immune detectors trained on the specific type of the attack (Fig. 4).

A compressed input data set of the dimension of 12 flows on the neural network immune detectors, each of them is trained on a specific type of the attack. If the detector identifies the attack then the output value of its first operation will be defined as the unit value. In [60] a technique was proposed to resolve conflicts by such classifier when several detectors are installed in a single state.

The statistical evaluation of the reliability for the designed classifier is performed using the ROC-analysis [61]. Let's consider the results of the neural immune detector activity trained on DoS_neptune attack (Table 1). To train the one detector the data set is used, consisting of 64 connections from one of four classes of network attacks (that is, 80% of all training data for the neural network), and 16 connections belonging to the class of legitimate network connections (representing 20% of the training sample). This relation of classes in the training set was obtained by experiment and showed the best results.

ROC-curvesshow the detector's ability to detect and classify certain types of network attacks (Fig. 5).

As it can be seen from the Table 1 and Fig. 5, a probability of the correct detection and classification for dos_neptune is 100% where the neural network immune detector was trained. In addition, this neural network immune detector detects and classifies the novel types of attacks with the different probability: dos_land attacks - 100%, probe_portsweep - 98,9%, probe_satan - 90,0%, r2l_imap - 66,7%. The probability of the correct detection and classification for normal connections is 99.9%, and,

Table 1. Detection and classification of network attacks by detectors trained on DoS_neptune

Type of attack	*Se (TPR),%*	*FNR,%*	*Accu,%*
DoS-attacks			
back	0,0	100,0	50,0
land	100,0	0,0	100,0
neptune	100,0	0,0	100,0
pod	0,0	100,0	50,0
smurf	0,0	100,0	50,0
teardrop	3,7	96,3	51,8
Probe-attacks			
ipsweep	6,5	93,5	53,2
nmap	0,0	100,0	50,0
portsweep	98,9	1,1	99,4
satan	90,0	10,0	95,0
R2L-attacks			
ftp_write	0,0	100,0	50,0
guess_passwd	1,9	98,1	50,1
imap	66,7	33,3	83,3
multihop	0,0	100,0	50,0
phf	0,0	100,0	50,0
spy	0,0	100,0	50,0
warezclient	0,2	99,8	50,1
U2R-attacks			
buffer_overflow	0,0	100,0	50,0
loadmodule	0,0	100,0	50,0
perl	0,0	100,0	50,0
rootkit	0,0	100,0	50,0

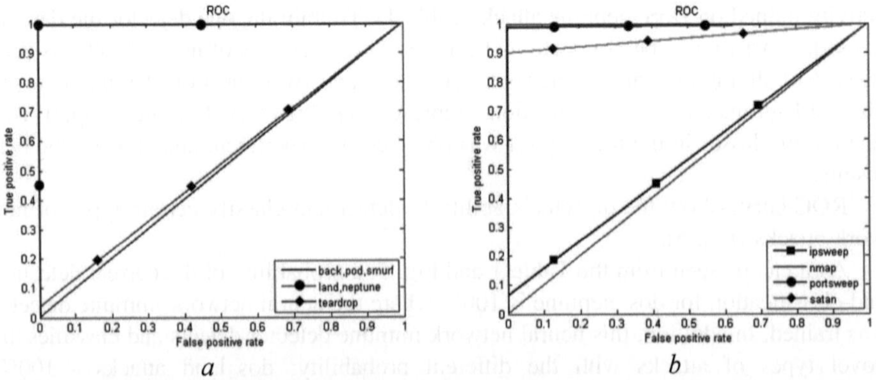

Fig. 5. ROC-curves show the detection and classification of attacks by neural network immune detector: *a* – DoS-attacks, *b* – Probe-attacks

correspondingly, the error of the second kind, characterizing the probability of false positives, is equal 0.1%.

5.2 Adaptation Ability of Cyber Defense System Detecting the Malware

The goal of this experiment is to show the adaptation ability of the proposed Artificial Cyber Defense System on the example of malware detection. Let us discuss briefly the experimental conditions. Initially, we generate several immune detectors, for example the five detectors D1…D5 (Table 2) which went through the both training and selection phase.

The learning set for each detector is unique and consists of data from randomly chosen legitimate files and the malicious code. The much-used method in the existing antivirus solutions can be employed for extracting the data from files. The main idea of this method is to define the common special chunks of executable files that are very special for the malicious software. A Table 3 represents the example of such areas and a Fig. 6 shows the example of the code for the learning set.

Table 2. Immune detectors

Detectors	Learning set
D1	eventvwr.exe, dllhost.exe, eventvwr.exe, fixmapi.exe, **Trojan-Downloader.Win32. Bagle.f**
D2	finger.exe, eventvwr.exe, loadfix.com, proxycfg.exe, **Email-Worm.Win32. Brontok.q**
D3	control.exe, proxycfg.exe, systray.exe, regwiz.exe, **Trojan-Proxy.Win32.Lager.d**
D4	forcedos.exe, rspndr.exe, share.exe, lpq.exe, **Net-Worm.Win32.Bozori.k**
D5	regedt32.exe, redir.exe, loadfix.com, control.exe, **Trojan-**

Table 3. The special areas of executable files

Offset	Name of area	Size
0x0000	Header	0x0400
0x0400	Page A	0x0400
0x0800	Page B	0x0400
0x0C00	Page E	0x0800
0x1400	Page C	0x4000

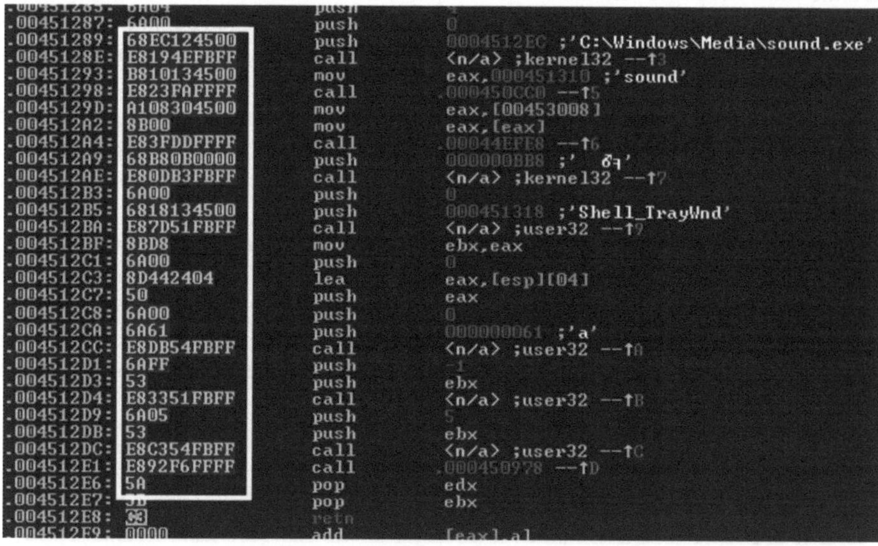

Fig. 6. The example of code for learning set

After training and selection neural network immune detectors check the set of malware. A Table 4 shows the detection ability of each immune detector (with the purpose of space saving Tables 2 and 4 are in the short form).

P_T and P_F are interdependent values which characterize the membership of the under-test file into legitimate or malicious class. P_T is the probability that checked object is legitimate. P_F is the inverse value which shows the belonging of the checked object to the malware class.

Values of P_T and P_F can be determined from equations:

$$P_T = \frac{\overline{Y_1}}{L},\ \overline{Y_1} = \sum_{k=1}^{L} Y_1^k,$$
$$P_F = 1 - P_T = \frac{\overline{Y_2}}{L},\ \overline{Y_2} = L - \overline{Y_2} = \sum_{k=1}^{L} Y_2^k, \tag{7}$$

where P_T is the probability of legitimate file; P_F is the probability of malware; $\overline{Y_1}$ and $\overline{Y_2}$ is the number of legitimate and malicious fragments of under-test file correspondingly; L is the total amount of fragments from under-test file, Y_{ik} is the i-th output of immune detector for k-th input pattern.

Fragments of files appear when we are dividing the checked file into the chunks with the size equals to the number of input neurons for neural network immune detectors. For example: if the file size equals 16 Kbyte and we use 128 input neurons in neural network immune detector then the number of fragments equals $L = 16*1024/128 = 128$.

We are using the threshold that defines the belonging of checking objects to class of malware. If $P_T > 0.8$ ($P_F < 0.2$ correspondingly) then the checked object is legitimate. Conversely if $P_T < 0.8$ ($P_F > 0.2$ correspondingly) then the checked file is malicious one.

The evolution of detectors is the important part of the intelligent security system, because it allows to expose new regularities and features of the novel malware and even to adapt to them.

Let's consider the process of evaluating the detectors in the proposed system. If i-th detector detects a malware, then it activates. In this case the evolution process occurs that contains both cloning and mutation procedures.

Table 4. The detection ability of detectors

Malware	D1,P_T/P_F	D2,P_T/P_F	D3,P_T/P_F	D4,P_T/P_F	D5,P_T/P_F
Worm.Brontok.q	**0,78/0,22**	0,83/0,17	0,83/0,17	0,85/0,15	**0,78/0,22**
Worm.NetSky.q	**0,74/0,26**	0,95/0,05	0,97/0,03	1,00/0,00	0,90/0,10
Worm.Rays	0,96/0,04	0,86/0,14	0,85/0,15	**0,79/0,21**	0,82/0,18
Worm.Bozori.a	**0,78/0,22**	0,93/0,07	0,95/0,05	0,99/0,01	0,88/0,12
Worm.Bozori.k	**0,77/0,23**	0,92/0,08	0,93/0,07	0,96/0,04	0,88/0,12
Packed.Tibs	**0,77/0,23**	0,96/0,04	0,97/0,03	0,99/0,01	0,92/0,08
Trojan.Dialer.eb	0,89/0,11	0,81/0,19	0,80/0,20	0,83/0,17	**0,79/0,21**
Trojan.INS.gi	0,86/0,14	**0,79/0,21**	**0,75/0,25**	**0,73/0,27**	**0,75/0,25**
Trojan.Small.dde	**0,77/0,23**	0,95/0,05	0,96/0,04	1,00/0,00	0,90/0,10
Trojan.Lager.d	0,83/0,17	0,88/0,12	0,75/0,25	0,93/0,07	**0,79/0,21**
Virus.Bee	0,97/0,03	**0,79/0,21**	**0,77/0,23**	**0,77/0,23**	0,80/0,20
Virus.Neshta.a	0,90/0,10	**0,74/0,26**	**0,72/0,28**	**0,72/0,28**	**0,72/0,28**
Virus.VB.d	0,93/0,07	**0,69/0,31**	**0,65/0,35**	**0,65/0,35**	**0,69/0,31**

Table 5. The detection ability of clones

Malware	C1, P_T/P_F	C2, P_T/P_F	C3, P_T/P_F	C4, P_T/P_F
Trojan.Bagle.f	**0,79/0,21**	**0,78/0,22**	**0,79/0,21**	**0,77/0,23**
Trojan.INS.bl	**0,77/0,23**	**0,79/0,21**	**0,78/0,22**	**0,76/0,24**
Trojan.INS.gi	**0,64/0,36**	**0,66/0,34**	**0,65/0,35**	**0,68/0,42**
Trojan.Ladder.a	**0,75/0,25**	**0,76/0,24**	**0,77/0,23**	**0,77/0,23**
Trojan.Small.da	0,83/0,17	0,81/0,19	**0,79/0,21**	0,80/0,20
Trojan.Small.dde	0,84/0,16	**0,79/0,21**	0,85/0,15	0,80/0,20
Trojan.Small.dg	**0,77/0,23**	**0,77/0,23**	**0,78/0,22**	**0,76/0,24**
Trojan.Daemon.a	**0,73/0,27**	**0,79/0,21**	**0,78/0,22**	**0,74/0,26**
Trojan.Lager.d	**0,78/0,22**	**0,78/0,22**	**0,78/0,22**	**0,77/0,23**
Trojan.Mitglied.o	**0,78/0,22**	**0,77/0,23**	**0,77/0,23**	**0,78/0,22**
Trojan.Small.a	0,81/0,19	0,82/0,18	0,82/0,18	**0,78/0,22**
Virus.Bee	**0,70/0,30**	**0,72/0,28**	**0,71/0,29**	**0,74/0,26**
Virus.Neshta.a	**0,71/0,29**	**0,69/0,31**	**0,70/0,30**	**0,72/0,28**
Virus.VB.d	**0,78/0,22**	0,81/0,19	0,88/0,12	**0,79/0,21**

The algorithm of cloning can be represented as the following set of steps:

- Creating the copies (clones) of the detector that found the malicious code.
- Creation of the learning set based on data from found malware.
- Training of clones.
- Calculating the fitness of clones.

In practice it means that data form the detected malware is included to the learning set and clone *Ci* (in the fact it is a neural network) as well as the learning process is exposed. As a result we get the set of clones C_i, with improved capabilities. The fitness function is used to determine the acquired new detection ability of clones. As fitness the mean-squared error between input and output vectors for the immune detector can be used (Eq. 5).

The cloning process is the important part of the proposed system because it helps detectors to adapt to the novel threat and provide the effective defense from the active cyber attack. The process of retraining of immune detectors, when data from detected malware are used, enables immune detectors to improve the detection ability, increase the detection rate and permit the whole system to adapt to the changeable environment by evolution means.

A Table 5 demonstrates the ability of detectors to recognize the novel malware.

The gray cells (see Tables 2 and 4) indicate the detection of malware. As can be seen from Tables 2, 4 and 5) each clone *Ci* of the "parent" detector D3 is capable to detect kinds of the known malware (in the Table 5 they are highlighted by the light gray color) as well as kinds of the novel malware (in the Table 5 they are highlighted by the dark gray color). Thus the detected novel malware is added to the training data set in order to train the new neural network immune detectors. It enables obtaining the new detectors with the different structure, tuned into novel malware. These two mechanisms, namely cloning and training dataset modification, provide increasing the detection quality and adapting detectors to the novel malware.

6 Conclusion and Future Work

The Intelligent System for Cyber Defense is developed with the ability to detect known and novel cyber attacks. This system is characterized as the adaptive, evolutionary and self-organizing one. The ability of immune detectors with neural network structure to adapt to changeable cyber-attacks trend and as result to evolve with the scope increasing the rate of novel cyber attacks is examined. We have run experiments proofing that detectors can adapt to the novel threat.

The adapted detectors acquire the ability to detect some novel attacks with a higher quality. Each novel detected attack is adding to the training sample that increases the difference in immune detectors and enables detecting novel attacks.

It is experimentally confirmed, that proposed neural network immune detectors are able to discover the novel types of attacks and adapt to them.

The proposed Intelligent Cyber Defense System can increase the reliability of intrusion detection in computer systems, and therefore it may reduce financial losses of companies from cyber attacks.

It's expected in the near future to implement a part of the Cyber Defense System units using the programmable logic arrays. Such solution, unlike the software protection, will allow eliminating the influence of the software intrusions on the Cyber Defense System.

Acknowledgements. This work is running under a grant by the Ministry of Education and Sciences, Ukraine, 2016–2017 as well as it's supported by the Belarusian State Research Program "Informatics and Space", 2011–2015.

References

1. Center for Internet Security. http://cisecurity.org
2. Cybercrime in the world. http://www.tadviser.ru/index.php (in Russian)
3. Symantec it estimated the annual losses from cybercrime at $ 114 billion. http://www.companion.ua/articles/content?id=162264 (in Russian)
4. Forbes experts have chosen the most high-profile cyber attacks recently. http://www.securitylab.ru/news/444700.php (in Russian)
5. Unuchek, R., Garnaeva, M., Makrushin, D., Sinitsyn, F., Liskin, A.: IT threat evolution Q3 2016. Statistics. https://securelist.com/analysis/quarterly-malware-reports/76513/it-threat-evolution-q3-2016-statistics
6. Worm.Win32.Stuxnet. https://threats.kaspersky.com/en/threat/Worm.Win32.Stuxnet
7. Shiels, M.: Cyber-sabotage and espionage top 2011 security fears. http://www.bbc.co.uk/news/technology-12056594
8. W32.Koobface. http://www.symantec.com/security_response/writeup.jsp?docid=2008-080315-0217-99

9. IBM X-Force 2012: Trend and Risk Report. https://www.ibm.com/ibm/files/I218646H25649F77/Risk_Report.pdf
10. Naraine, R.: Duqu FAQ. http://www.securelist.com/en/blog/208193178/Duqu_FAQ
11. Kaspersky Security Bulletin. Malware Evolution 2011 (2011). http://securelist.com/analysis/kaspersky-security-bulletin/36494/kaspersky-security-bulletin-malware-evolution-2011/
12. KDD Cup 1999 Data (1999). http://kdd.ics.uci.edu/databases/kddcup99/kddcup99.html
13. Tavallaee, M., Bagheri, E., Lu, W., et al.: A detailed analysis of the KDD CUP 99 data set. In: Proceedings of the IEEE Symposium on Computational Intelligence in Security and Defense Applications (CISDA 2009), pp. 1–8 (2009). doi:10.1109/CISDA.2009.5356528
14. Lia, L.B., Chang, R.I., Kouh, J.S.: Detecting network intrusions using signal processing with query-based sampling filter. EURASIP J. Adv. Sig. Process., Article ID 735283, 1–8 (2009). Hindawi Publishing Corporation
15. Laheeb, M.I.: Anomaly network intrusion detection system based on distributed time-delay neural network. J. Eng. Sci. Technol. **5**(4), 457–471 (2010)
16. Cannady, J.: Artificial neural networks for misuse detection. In: Proceedings of the 21st National Information Systems Security Conference, Arlington, VA, USA, 5–8 October 1998, pp. 368–381 (1998)
17. Chen, W.H., Hsu, S.H., Shen, H.P.: Application of SVM and ANN for intrusion detection. Comput. Oper. Res. **32**(10), 2617–2634 (2005)
18. Mukkamalaa, S., Sung, A.H., Abraham, A.: Intrusion detection using an ensemble of intelligent paradigms. J. Network Comput. Appl. **28**(2), 167–182 (2005)
19. Lorenzo-Fonseca, I., Maciá-Pérez, F., Mora-Gimeno, F.J., Lau-Fernández, R., Gil-Martínez-Abarca, J.A., Marcos-Jorquera, D.: Intrusion detection method using neural networks based on the reduction of characteristics. In: Cabestany, J., Sandoval, F., Prieto, A., Corchado, J.M. (eds.) IWANN 2009. LNCS, vol. 5517, pp. 1296–1303. Springer, Heidelberg (2009). doi:10.1007/978-3-642-02478-8_162
20. Kang, B.-D., Lee, J.-W., Kim, J.-H., Kwon, O.-H., Seong, C.-Y., Park, S.-M., Kim, S.-K.: A mutated intrusion detection system using principal component analysis and time delay neural network. In: Wang, J., Yi, Z., Zurada, J.M., Lu, B.-L., Yin, H. (eds.) ISNN 2006. LNCS, vol. 3973, pp. 246–254. Springer, Heidelberg (2006). doi:10.1007/11760191_36
21. Grediaga, Á., Ibarra, F., García, F., Ledesma, B., Brotóns, F.: Application of neural networks in network control and information security. In: Wang, J., Yi, Z., Zurada, J.M., Lu, B.-L., Yin, H. (eds.) ISNN 2006. LNCS, vol. 3973, pp. 208–213. Springer, Heidelberg (2006). doi:10.1007/11760191_31
22. Zhang, C., Jiang, J., Kamel, M.: Comparison of BPL and RBF network in intrusion detection system. In: Wang, G., Liu, Q., Yao, Y., Skowron, A. (eds.) RSFDGrC 2003. LNCS, vol. 2639, pp. 466–470. Springer, Heidelberg (2003). doi:10.1007/3-540-39205-X_79
23. Kohonen, T.: The self organizing map. Proc. Inst. Electr. Electronics Eng. **78**, 1464–1480 (1990)
24. Cannady, J.: Applying CMAC-based online learning to intrusion detection. In: Proceedings of the International Joint Conference on Neural Networks, (IJCNN 2000), IEEE-INNS-ENNS, vol. 5, pp. 405–410 (2000)
25. Debar, H., Becker, M., Siboni, D.: A neural network component for an intrusion detection system. In: Proceedings of the IEEE Computer Society Symposium on Research in Security and Privacy, pp. 240–250 (1992)
26. Cheng, E., Jin, H., Han, Z., Sun, J.: Network-based anomaly detection using an elman network. In: Lu, X., Zhao, W. (eds.) ICCNMC 2005. LNCS, vol. 3619, pp. 471–480. Springer, Heidelberg (2005). doi:10.1007/11534310_51

27. Höglund, A.J., Hätönen, K., Sorvari, A.S.: A computer host-based user anomaly detection system using the self-organizing map. In: Proceedings of the IEEE-INNS-ENNS International Joint Conference on Neural Networks (IJCNN 2000), vol. 5, pp. 411–416 (2000)
28. Ramadas, M., Ostermann, S., Tjaden, B.: Detecting anomalous network traffic with self-organizing maps. In: Vigna, G., Kruegel, C., Jonsson, E. (eds.) RAID 2003. LNCS, vol. 2820, pp. 36–54. Springer, Heidelberg (2003). doi:10.1007/978-3-540-45248-5_3
29. Sarasamma, S.T., Zhu, Q.A., Huff, J.: Hierarchical Kohonen net for anomaly detection in network security. IEEE Trans. Syst. Man Cybern. Part B **35**(2), 302–312 (2005)
30. Jirapummin, C., Wattanapongsakorn, N., Kanthamanon, P.: Hybrid neural networks for intrusion detection system. In: Proceedings of the International Technical Conference on Circuits/Systems, Computers and Communications, Thailand, pp. 928–931 (2002)
31. Horeis, T.: Intrusion detection with neural networks – Combination of self-organizing maps and radial basis function networks for human expert integration, Technical report. http://citeseerx.ist.psu.edu/viewdoc/download?doi=10.1.1.106.191&rep=rep1&type=pdf
32. Chimphlee, W., Abdullah, A.H., Sap, M.N.M.: Anomaly-based intrusion detection using fuzzy rough clustering. In: Proceedings of the International Conference in Hybrid Information Technology (ICHIT 2006), vol. 1, pp. 329–334 (2006)
33. Dickerson, J.E., Juslin, J., Koukousoula, J., Dickerson, J.A. Fuzzy intrusion detection. In: Proceedings of the 20th International Conference of the North American Fuzzy Information Society (NAFIPS 2001) and Joint the 9th IFSA World Congress, 3, Vancouver, Canada, vol. 3, pp. 1506–1510 (2001)
34. Forrest, S., Perelson, A., Allen, L., Cherukuri, R.: Self-nonself discrimination in a computer. In: Proceedings of the IEEE Symposium on Security and Privacy, pp. 202–212 (1994)
35. Forrest, S., Hofmeyr, S., Somayaji, A.: Computer Immunology. Commun. ACM **40**(10), 88–96 (1997)
36. Hofmeyr, S., Forrest, S.: Immunity by design. In: Proceeding of the Genetic and Evolutionary Computation Conference (GECCO 1999), pp. 1289–1296 (1999)
37. Burnet, F.: The Clonal Selection Theory of Acquired Immunity. Cambridge University Press, Cambridge (1959)
38. Burnet, F.: Clonal selection and after. In: Bell, G., Perelson, A., Pimbley, G. (eds.) Theoretical Immunology, pp. 63–85. Marcel Dekker Inc., New York (1978)
39. Jerne, N.: Towards a network theory of the immune system. Ann. Immunology (Inst. Pasteur) **125C**, 373–389 (1974)
40. Greensmith, J., Whitbrook, A., Aickelin, U.: Artificial immune systems. In: Handbook of Metaheuristics, 2nd edn., vol. 14, pp. 421–448. Springer, New York (2010)
41. Greensmith, J., Aickelin, U.: The deterministic dendritic cell algorithm. In: Bentley, P.J., Lee, D., Jung, S. (eds.) ICARIS 2008. LNCS, vol. 5132, pp. 291–302. Springer, Heidelberg (2008). doi:10.1007/978-3-540-85072-4_26
42. Coico, R., Sunshine, G., Benjamini, E.: Immunology: A Short Course. Wiley-Liss (2003)
43. Murphy, K., Travers, P., Walport, M.: Janeway's Immunobiology. Garland Science, 7th edn. (2008)
44. Stibor, T., Mohr, P., Timmis, J., Eckert, C.: Is negative selection appropriate for anomaly detection? In: Proceedings of the Genetic and Evolutionary Computation Conference (GECCO 2005), pp. 321–328. Springer (2005)
45. Harmer, P., Williams, P., Gunsch, G., Lamont, G.: An artificial immune system architecture for computer security applications. IEEE Trans. Evol. Comput. **6**(3), 252–280 (2002)
46. Balthrop, J., Esponda, F., Forrest, S., Glickman, M.: Coverage and generalization in an artificial immune system. In: Proceedings of the Genetic and Evolutionary Computation Conference (GECCO 2002), pp. 3–10 (2002)

47. Stibor, T., Bayarou, K., Eckert, C.: An investigation of r-chunk detector generation on higher alphabets. In: Proceedings of the Genetic and Evolutionary Computation Conference (GECCO 2004), pp. 299–307 (2004)
48. Gonzales, F., Dasgupta, D., Gomez, J.: The effect of binary matching rules in negative selection. In: Proceedings of the Genetic and Evolutionary Computation Conference (GECCO 2003), pp. 198–209 (2003)
49. Kim, J., Bentley, P., Aickelin, U., et al.: Immune system approaches to intrusion detection – A review. Natural Comput. **6**(4), 413–466 (2007). Springer
50. Haykin, S.: Neural Networks: A Comprehensive Foundation, p. 842. Prentice Hall, Upper Saddle River (1999)
51. Golovko, V., Bezobrazov, S., Kachurka, P., Vaitsekhovich, L.: Neural network and artificial immune systems for malware and network intrusion detection. In: Koronacki, J., Raś, Z.W., Wierzchoń, S.T., Kacprzyk, J. (eds.) Advances in Machine Learning II. SCI, vol. 263, pp. 485–513. Springer, Heidelberg (2010)
52. Komar, M., Golovko, V., Sachenko, A., Bezobrazov, S.: Development of neural network immune detectors for computer attacks recognition and classification. In: Proceedings of the 7th IEEE International Conference on Intelligent Data Acquisition and Advanced Computing Systems: Technology and Applications (IDAACS-2013), Berlin, Germany, vol. 2, pp. 665–668 (2013)
53. Komar, M., Sachenko, A., Golovko, V., Bezobrazov, S.: Method of detection of computer attacks by the neural network artificial immune system, Pat. Number 109640 Ukraine (2015) (in Ukrainian)
54. Komar, M., Sachenko, A., Bezobrazov, S., Golovko, V., Intelligent cyber defense system. In: Proceedings of the 12th International Conference on ICT in Education, Research and Industrial Applications. Integration, Harmonization and Knowledge Transfer (ICTERI 2016), Kyiv, Ukraine, 21–24 June, pp. 534–549. CEUR-WS.org (2016)
55. Scholz, M., Fraunholz, M., Selbig, J.: Nonlinear principal component analysis: neural network models and applications. In: Gorban, A.N., Kégl, B., Wunsch, D.C., Zinovyev, A. Y. (eds.) Principal Manifolds for Data Visualization and Dimension Reduction, pp. 44–67. Springer, Heidelberg (2008)
56. Komar, M., Golovko, V., Sachenko, A., Bezobrazov S.: Intelligent system for detection of networking intrusion. In: Proceedings of the 6th IEEE International Conference on Intelligent Data Acquisition and Advanced Computing Systems: Technology and Applications (IDAACS-2011), Prague, vol. 1, pp. 74–377. Czech Republic (2011)
57. Hinton, G., Osindero, S., Teh, Y.: A fast learning algorithm for deep belief nets. Neural Comput. **18**, 1527–1554 (2006)
58. Bengio, Y.: Learning deep architectures for AI. Found. Trends Mach. Learn. **2**(1), 1–127 (2009)
59. Golovko, V., Kroshchanka, A., Rubanau, U., Jankowski, S.: A learning technique for deep belief neural networks. In: Golovko, V., Imada, A. (eds.) ICNNAI 2014. CCIS, vol. 440, pp. 136–146. Springer, Cham (2014). doi:10.1007/978-3-319-08201-1_13
60. Sachenko A., Komar M.: Intrusion detection system based on neural networks. Zeszyty Naukowe. Organizacja i Zarządzanie, Politechnika Śląska **68**, 377–386 (2014)
61. Fawcett, T.: Using rule sets to maximize ROC Performance. In: Proceedings of the IEEE International Conference on Data Mining (ICDM–2001), Los Alamitos, pp. 131–138 (2001)

Application of Ordered Fuzzy Decision Trees in Construction of Structure Function of Multi-State System

Elena Zaitseva(✉), Vitaly Levashenko, Miroslav Kvassay, and Jan Rabcan

Department of Infromatics,
University of Zilina, Univerzitna 8215/1, 010 26 Zilina, Slovakia
{elena.zaitseva,vitaly.levashenko,miroslav.kvassay, jan.rabcan}@fri.uniza.sk

Abstract. The structure function is typical mathematical representation of investigated system in reliability analysis. This function defines the correlation of all possible system components states and system performance level from point of view of the system reliability. The structure function is constructed based on complete information about the system structure and possible components states. However, there are a lot of practical problems when the complete information is not available because data from which it can be derived cannot be collected. In this paper, we propose a new method for construction of the structure function based on uncertain or incomplete initial data with application of ordered Fuzzy Decision Trees.

Keywords: Fuzzy decision tree · Multi-State System · Structure function · Uncertainty

1 Introduction

Two basic approaches for mathematical representation of investigated system depending on the number of system performance levels and its components states are used in reliability analysis. According to the first of them, the investigated system and its components have two possible levels: perfect functioning and complete failure. Such system representation is named as a *Binary-State System* (BSS). However, in practice, many systems can go through different performance levels between these two extreme states [1, 2]. A *Multi-State Systems* (MSS) is a mathematical model that is used to describe such systems since it allows defining more than two levels of performance [2–4].

Different mathematical models are used for BSSs and MSSs. One of them is structure function. In this case, the system is modeled as a mapping that assigns system performance levels to all possible combinations of component states. There are a lot of methods and algorithms for the calculation of different indices and measures for the reliability evaluation based on structure function as mathematical model of investigated system [1, 2, 4]. The structure function definition supposes the exact definition of all possible states of the system and its components. Therefore, the structure function

A. Ginige et al. (Eds.): ICTERI 2016, CCIS 783, pp. 56–75, 2017.
https://doi.org/10.1007/978-3-319-69965-3_4

cannot be defined if initial data about the system is uncertain and/or incomplete. It is important restriction to use the structure function for mathematical representation of real-world system, because data about behavior of such systems is uncertain and incomplete. Therefore, special methods have to be developed for the structure function construction based on uncertain and incomplete data.

A new approach for construction of the structure function based on uncertain and incomplete data has been proposed in [5, 6]. Methods of Data Mining for analysis of uncertain and incomplete data have been used in this approach. In particular, the structure function construction has been proposed based on the application of *Fuzzy Decision Trees* (FDT). FDTs are widely used in Data Mining for analysis of uncertain data and decision making with ambiguities. In this case, collected data for structure function construction can be defined with possibility or confidence. In addition, FDTs allow taking into account uncertainties caused by incomplete specified data. This is possible when it is expensive to obtain all data about real system behavior or there is little data with poor documentation. As a rule, if the exact value of the actual data about the system behavior cannot be determined, we need to rely on more data to give additional information necessary to correct the theoretical model [5, 6]. An FDT allows reconstructing these data with different levels of the possibility (confidence).

Many methods for constructing FDTs of different types have been developed. In [5, 6], a non-ordered FDT has been used for the structure function construction. Experimental investigation implemented and presented in [5, 6] has shown this FDT is effective for the structure function construction based on uncertain and incomplete data. But the non-ordered FDT has not some special properties that correlate with specifics of initial data or conditions of decided problem [7]. A choice of other type of an FDT can influence the result because, in case of the structure function construction, the modification of FDT type correlates with the error rate for the reconstructed structure function.

In this paper, the method for the structure function construction by application of an FDT is developed and a new type of FDT (ordered FDT) for construction of this function is used. This type of an FDT supposes the parallel procedure in some attributes classification that can be important property for evaluation of real-world systems.

Ordered FDTs have been proposed in [7, 8]. They permit finding a sequence of rules that analyze input attributes in order to ensure both cost effective and a desired level of accuracy. Every node of one level of such FDT associates with similar attribute. The ordered FDT induction has been developed based on cumulative information estimates of fuzzy sets in [7, 8] too. The cumulative information estimation is also used in this paper for the ordered FDT induction. The definition of cumulative information estimation in papers [5, 6] is adapted and developed for the structure function construction in this paper.

In this paper, the new type of FDTs is used for the structure function construction based on uncertain and incompletely specified data about investigated system. This type is ordered FDT that has the same attributes in each level of the tree. Such structure allows using parallel procedure in analyses of some attributes. The efficiency of the ordered FDT as error rate is analyzed and compared with previous investigations based on application of non-ordered FDT [5, 6].

This paper is structured as follows. Section 2 discusses the concept of the structure function and specifics of uncertainty of initial data in reliability analysis of real-world systems. The principal steps of the proposed method are considered in Sect. 3. These steps are: the collection of data into a repository (Sect. 3.1), the representation of the system model in the form of an FDT that is implemented by the ordered FDT induction (Sect. 3.2), and the construction of the structure function based on the FDT (Sect. 3.3). The illustration of the proposed method for an offshore electrical power generation system is in Sect. 4. This system was analyzed in paper [6] by the similar method with application of a non-ordered FDT. The comparison of accuracy and complexity of two methods (with application of non-ordered and ordered FDTs) are presented in Sect. 4 too.

2 Mathematical Representation of Multi-State System

A MSS is a mathematical model for investigated system/object representation in reliability engineering that allows indicating M performance levels from complete failure to perfect work in mathematical description of the investigated system/object. As a rule, the complete failure (the worst performance level) is interpreted as 0 and perfect work (the best performance level) agrees with the value $M - 1$ in mathematical description. The MSS consists of n components that can be in one of several (not only two) states. The component fault is presented as 0 and its best state is indicated as $m_i - 1$ in mathematical description ($i = 1,2,\ldots, n$).

Several methods for mathematical description and evaluation of a MSS [1–3] exist. One of them is representation of the MSS by structure function that maps possible components states to system performance levels [1]. Important advantage of the structure function is a possibility to describe system of any structural complexity.

2.1 Structure Function

The MSS performance levels are unambiguously determined by the states of components. Let us assume the investigated system is in the stationary state or evaluated in fixed time. Let the i-th component x_i have m_i states, i.e. states $\{0, 1,\ldots, m_i - 1\}$, and the space of the possible combinations of all components is $L^n = \{0,1,\ldots,m_1 - 1\} \times \ldots \times \{0,1,\ldots,m_n - 1\}$, and let $\{0, 1, \ldots, M - 1\}$ be the space of possible values of the performance levels of the entire system. The transformation $\phi(\boldsymbol{x})$:

$$\phi(x) = \phi(x_1,\ldots,x_n) : \{0,\ldots,m_1 - 1\} \times \ldots \times \{0,\ldots,m_n - 1\} \rightarrow \{0,\ldots,M - 1\} \quad (1)$$

which maps the space of the components' states into the space of the system's performance levels, is named as the structure function [1].

The structure function (1) can be interpreted as a classification procedure as shown in [5] possible combinations of all components states L^n are classified into M groups that agree with number of the system performance levels. For example, let us consider the system of parallel structure that consists of two components (n = 2). The first of them has 2 states (m_1 = 2), the second component has 4 states (m_2 = 4), and the system has 3 performance levels ($M = 3$). The structure function of this system is shown in Fig. 1. The system components states are classified into 3 groups: system failure, partly working states and working states.

The table of the structure function

x_1 x_2	$\phi(\boldsymbol{x})$
0 0	0
0 1	1
0 2	1
0 3	1
1 0	1
1 1	2
1 2	2
1 3	2

$\Rightarrow$

The classification of system states based on the structure function

$\phi(\boldsymbol{x})$		
0	1	2
(0 0)	(0 1)	(1 1)
	(0 2)	(1 2)
	(0 3)	(1 3)
	(1 0)	

Fig. 1. The structure function of the parallel system

Depending on the system type, the structure function (1) can have some specific properties. In this paper, coherent systems will be considered. This means: (a) the system structure function is monotone: $\phi(x_i, \boldsymbol{x}) \leq \phi(x_j, \boldsymbol{x})$ for any $x_i \leq x_j$; and (b) there are not irrelevant components in the system.

The evaluation of the MSS based on the structure function supposes knowledge of the probabilities of individual states for every system component:

$$p_{i,s} = \Pr\{x_i = s\}, s = 0, \ldots, m_i - 1. \tag{2}$$

A lot of reliability indices and measures can be calculated based on the structure function. The structure function (1) allows calculating the boundary system states [3], minimal cut/path sets [2, 4] and importance measures [9]. One of basic indices is the probability of system performance level that is calculated as [1]:

$$A_j = \Pr\{\phi(x) = j\}, j = 0, \ldots, m_i - 1. \tag{3}$$

However, identification of the structure function as Eq. (1) for a real application can be a difficult problem since this equation supposes unambiguously determination of the system performance level depending on all possible components states. The structure function can be defined as a result of the system structure analysis or based on expert data [10, 11]. In system structure analysis, the system is interpreted as a set of components (subsystems) with correlations. These correlations can be defined by functional relations that are interpreted as the structure function (1). However, there are many structure-complex systems for which correlations and/or connections of components are hidden or uncertain (e.g. power systems, network systems). The structure function in this case is incompletely specified and/or ambiguous. The expert's data for the structure function construction is ambiguous too [10, 12]. So, the problem of construction of the structure function (1) based on ambiguous and uncertain data is important. In this paper, we develop a new method for the structure function construction based on ambiguous and uncertain data, but specifics of uncertainty in reliability analysis have to be researched before the method consideration.

2.2 Uncertainty of Data

There are different factors that caused uncertainty of data collected for reliability analysis. In our investigation, we will take into account two of them. The first are ambiguity and vagueness of initial data. The second is incompleteness of data.

The ambiguity and vagueness of initial data for reliability analysis are caused by methods for the data collection: measurement and/or expert's knowledge. The measurement can be inaccurate and with an error that depends on accuracy of measuring device. Therefore these data can be defined and used with some likelihood. In case of data collected based on expert's knowledge, data cannot be indicate exact, because experts can have different opinions on one situation. Therefore, values of states of the components or system performance level cannot be indicated as exact (integers). Ambiguity and vagueness in a real system have been studied using the probability theory. However, it is worth pointing out that some uncertainties that are not random in nature can play important roles in construction of the structure function [10–13]. The fuzzy logic makes it possible to define the structure function in a more flexible form for such data than the probabilistic approach. So, non-exact values are the first factor of the uncertainty of initial data, and it can be expressed using fuzzy values.

The incompleteness of data depends on situations in which it is impossible to indicate some values of the system components states or performance level can exit. For example, it can be very expensive, or it needs unacceptable long time, or it is dangerous for the environment. This implies that some information about the system behavior can be absent and space of possible components states is incomplete. Therefore, initial data for the structure function construction is uncertain and incompletely specified. But the structure function according to (1) is defined for all possible states of the system components. It means that the exact and completely specified structure function has to be constructed based on uncertain data.

Let us to take into account the classification context of the structure function (Fig. 1). In this case, the problem of the structure function (1) construction can be interpreted as the development of classification method based on uncertain and incomplete data. It is a typical problem of Data Mining [14]. One of the approaches used for solving this problem is application of *Fuzzy Decision Trees* (FDTs), which are widely used in Data Mining for analysis of uncertain data and decision making in ambiguities [15, 16].

A decision tree is a formalism for expressing mappings of input attributes (components states) to output attribute/attributes (system performance level), consisting of an analysis of attribute nodes (input attributes) linked to two or more sub-trees and leafs or decision nodes labeled with classes of the output attribute (in our case, a class agrees with a system performance level) [17]. An FDT is one of the possible types of decision trees that permit operating with fuzzy data (attributes) and that use methods of fuzzy logic.

An FDT is a classification structure and its induction is possible based on incompletely specified data. FDT allows us to take into account uncertainty of initial data by the use of fuzzy values for components states (structure function's variables) and system performance level (structure function's value). An FDT can be inducted based on uncertain and incomplete data using some (not all) samples. A sample consists of some input and output attributes: the state vector $\boldsymbol{x} = (x_1,\ldots, x_n)$ is interpreted as input attributes and the value of the structure function as output attribute.

Inducted FDT permits classifying known as well as new samples. Samples are components states for the structure function induction. Therefore, the structure function can be constructed for all possible combinations of components states if they are defined as samples for FDT. In terms of Data Mining, the structure function is interpreted as a table of decisions [5, 10].

The structure function is constructed as a decision table that classifies the system performance level for each possible combination of components states. The decision table is formed based on the FDT that provides the mapping for all possible components states (input data) to M performance levels. This correlation of decision table and structure function has been used in the methodology for the structure function construction based on uncertain data.

3 The Structure Function Construction Based on Uncertain Data with Application of Fuzzy Decision Tree

The methodological background of the structure function construction based on uncertain data has been introduced in papers [5, 6]. The general procedure for the structure function construction includes next steps (Fig. 2):

- collection of data into the repository according to requests of FDT induction;
- representation of the system model in the form of an FDT that classifies components states according to the system performance levels;
- construction of the structure function as a decision table that is created by inducted FDT.

This general procedure is used for the structure function construction by the ordered FDT too. The principal difference is the second step where the ordered FDT is used for the representation of the classification model of the structure function based on uncertain and incomplete initial data.

3.1 Data Repository

Initial data for the structure function construction is collected in a form of the repository by the monitoring or expert evaluation of values of system component states and system performance levels. The repository for the FDT induction is formed as a table. The number of columns is $n + 1$. The first n columns are used for the representation of

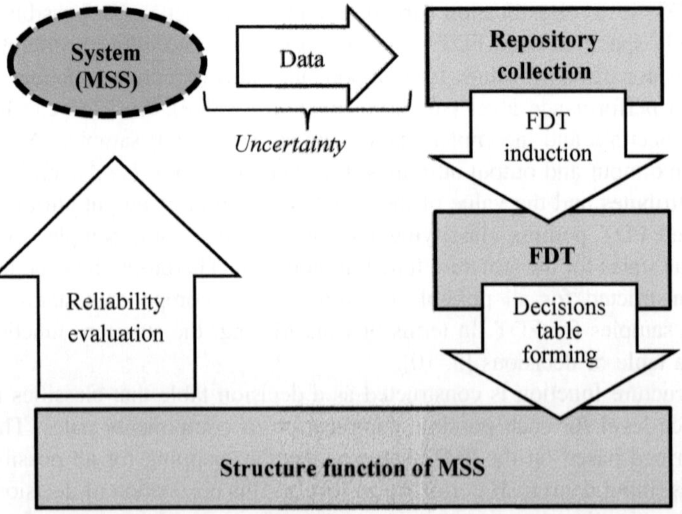

Fig. 2. The structure function induction based on data mining methods

states of the components. The system performance level is indicated in the last column. The i-th column in the first n columns includes m_i sub-columns (i = 1, 2,…, n). The last column for the system performance level has M sub-columns. The sub-column is linked with one of the values of component states or performance levels.

Every row of the table represents one monitoring situation or one expert's evaluation. The number of rows agrees with the number of samples for FDT induction.

The table's cell contains a number (from interval from 0 to 1) that is interpreted as the possibility (likelihood) of the value (from 0 to $m_i - 1$ or $M - 1$) indicated in the sub-column in which this cell is. Note that the sum of these possibilities in cells of one column and one row equals to 1. It means that the sum is 1 for possibilities of all states of every system component or system performance levels in the sample. Such data can be obtained from expert evaluations or by possibilistic fuzzy clustering [18]. These possibilities correspond to a membership function of fuzzy data [19]. This demand for initial data representation is caused by the method of FDT induction.

3.2 FDT Induction

There are different types of FDT. An FDT named as non-ordered has been used in [5, 6] for the structure function construction based on uncertain data. Ordered FDT has been introduced in [7]. This FDT has the same attributes in every level. It permits to use the predefined order of attributes in analysis of new samples: the value of next attribute can be defined before the computation of result of current level. In this paper the ordered FDT is used for the construction of the structure function based on uncertain data. The method for the ordered FDT induction with application of cumulative information estimates has been proposed in [7]. This method is used for induction of ordered FDT for construction of the structure function in this paper.

The induction of an FDT of any type is performed by the determination of the correlation between n input attributes $\{A_1, A_2,\ldots, A_n\}$ and one output attribute B. The construction of the system structure function supposes that the system performance level is the output attribute and component states defined by a state vector are input attributes [5]. The correlation between the terminologies and basic concepts of FDTs and reliability analysis was introduced in [5, 6]. Let us to consider this correlation succinctly.

Input attribute (component state) A_i ($i = 1, 2,\ldots, n$) corresponds to the i-th column of repository and is measured by a group of discrete values ranging from 0 to $m_i - 1$. These values $\{A_{i,0},\ldots, A_{i,j},\ldots, A_{i,m_i-1}\}$ agree with the values of states of the i-th component and correspond to sub-columns of the i-th column in the repository. An FDT assumes that the input set $A = \{A_1, A_2,\ldots, A_n\}$ is classified into one of the values of output attribute B. Output attribute B corresponds to the last (the $n + 1$-th) column of the repository. Value B_w of output attribute B agrees with one of the system performance levels and is defined by M values ranging from 0 to $M - 1$ ($w = 0,1,\ldots, M - 1$).

A fuzzy set A with respect to a universe U is characterized by a *membership function* $\mu_A(u)$: $U \rightarrow [0,1]$ that assigns an A-membership degree, $\mu_A(u)$, to each element u in U. $\mu_A(u)$ gives us an estimation that u belongs to A. The cardinality measure of the fuzzy set A is defined by $M(A) = \sum_{u \in U} \mu_A(u)$, and it is a measure of size of set A. For $u \in U$, $\mu_A(u) = 1$ means that u is definitely a member of A while $\mu_A(u) = 0$ means that u is definitely not a member of A, while $0 < \mu_A(u) < 1$ means that u is a partial member of A. If either $\mu_A(u) = 0$ or $\mu_A(u) = 1$ for all $u \in U$, A is a crisp set.

The FDT induction is based on determination of expanded attributes A_{i_q}. The type of FDT is caused by the criteria for the choice of this attribute. The selection criterion of expanded attributes A_{i_q} for induction of a non-ordered FDT is defined as:

$$iq = \operatorname{argmax}\left(\frac{\mathbf{I}(B;\ A_{i_1},\ldots, A_{i_{q-1}},\ A_{i_q})}{Cost\ (A_{i_q}) \times \mathbf{H}(A_{i_q})}\right), \tag{4}$$

where $A_{i_1},\ldots, A_{i_{q-1}}$ are input attributes which have been used in the previous levels of FDT; A_{i_q} is the attribute that is not used yet; $\mathbf{I}(B;A_{i_1},\ldots, A_{i_{q-1}}, A_{i_q})$ is cumulative mutual information. $Cost(A_{i_q})$ is an integrated measure that covers financial and temporal costs required to define the value of the A_{i_q} for an instance and this value is defined a priori. $\mathbf{H}(A_{i_q})$ is a cumulative entropy of this input attribute A_{i_q}.

The cumulative mutual information in output attribute B about the attribute A_{i_q} and the sequence of attributes $\{A_{i_1},\ldots, A_{i_{q-1}}\}$ reflects the influence of attribute A_{i_q} on the output attribute B when sequence $\{A_{i_1},\ldots, A_{i_{q-1}}\}$ is known. This measure has been introduced in [7] and calculated as:

$$\begin{aligned}\mathbf{I}(B;\ A_{i_1},\ldots, A_{i_{q-1}},\ A_{i_q}) = \sum_{j=0}^{m_b-1} \sum_{j_q=0}^{m_{i_q}-1} \sum_{j_1=0}^{m_{i_1}-1} \cdots \sum_{j_{q-1}=0}^{m_{i_{q-1}}-1} M(B_j \times A_{i_1 j_1} \times \ldots \times A_{i_q j_q}) \times \\ \times \left(\mathbf{I}(B_j, U_{q-1}) + \mathbf{I}(U_{q-1}, A_{i_q j_q}) - \mathbf{I}(B_j, U_{q-1}, A_{i_q j_q}) - \mathbf{I}(U_{q-1})\right)\end{aligned} \text{ bits,} \tag{5}$$

where $M(B_j \times A_{i_1,j_1} \times \ldots \times A_{i_q,j_q})$ is a cardinality measure of fuzzy set $B_j \times A_{i_1,j_1} \times \ldots \times A_{i_q,j_q}$; $U_{q-1} = \{A_{i_1,j_1}, \ldots, A_{i_{q-1},j_{q-1}}\}$ is a sequence of attribute's values; $\mathbf{I}(B_j, U_{q-1})$, $\mathbf{I}(U_{q-1}, A_{i_q,j_q})$, $\mathbf{I}(B_j, U_{q-1}, A_{i_q,j_q})$ and $\mathbf{I}(U_{q-1})$ are cumulative joint information.

Note, this cumulative mutual information equals the difference between cumulative conditional entropies:

$$\mathbf{I}(B; A_{i_1}, \ldots, A_{i_{q-1}}, A_{i_q}) = \mathbf{H}(B| A_{i_1}, \ldots, A_{i_{q-1}}) - \mathbf{H}(B| A_{i_1}, \ldots, A_{i_{q-1}}, A_{i_q}).$$

The entropy $\mathbf{H}(B| A_{i_1}, \ldots, A_{i_{q-1}})$ describes the uncertainty of a situation if previous attributes $A_{i_1}, \ldots, A_{i_{q-1}}$ are known and attribute A_{i_q} is unknown. The next cumulative conditional entropy $\mathbf{H}(B| A_{i_1}, \ldots, A_{i_{q-1}}, A_{i_q})$ describes a situation when attributes $A_{i_1}, \ldots, A_{i_{q-1}}$ and A_{i_q} are known.

Let us have a sequence of $q - 1$ input attributes $A_{i_1}, \ldots, A_{i_{q-1}}$ and one output attribute B. The cumulative joint information of the sequence of values $U_{q-1} = \{A_{i_1,j_1}, \ldots, A_{i_{q-1},j_{q-1}}\}$ ($q \geq 2$) and value B_j is:

$$\mathbf{I}(B_j, U_{q-1}) = \log_2 N - \log_2 M(B_j \times A_{i_1,j_1} \times \ldots \times A_{i_{q-1},j_{q-1}}) \text{ bits.} \quad (6)$$

The cumulative entropy of input attribute A_{i_q} reflects the ambiguity of this attribute. We calculate this entropy using the next rule:

$$\mathbf{H}(A_{i_q}) = \sum_{j_q=0}^{m_{i_q}-1} M(A_{i_q,j_q}) \times \mathbf{I}(A_{i_q,j_q}) \text{ bits,} \quad (7)$$

where $M(A_{i_q,j_q})$ is the cardinality measure of set A_{i_q,j_q}, and $\mathbf{I}(A_{i_q,j_q})$ is the cumulative joint information.

Maximum value i_q in (4) facilitates the selection of expanded attribute A_{i_q}. This attribute will be associated with a node of the FDT.

There are two tuning thresholds α and β in this method of FDT induction [7, 20]. A tree branch stops to expand when either the frequency *f* of the branch is below α or when more than β percent of instances left in the branch has the same class label. These values are thus key parameters needed to decide whether we have already arrived at a leaf node or whether the branch should be expanded further. Decreasing the parameter α and increasing the parameter β allow us to build large FDTs. On one hand, large FDTs describe datasets in more detail. On the other hand, these FDTs are very sensitive to noise in the dataset. We empirically selected parameters near α = 0.10 and β = 0.90. We estimated that a confidence degree of more than 0.90 would allow us to reach a decision with sufficient confidence. Moreover, the threshold frequency 0.10 eliminates the variants of no-principal decisions. Notably, increasing the size of the FDT has no influence on the FDT root or the higher FDT nodes. It only adds new nodes and leaves to the bottom part of the FDT. These new nodes and leaves have a low bearing on decision making. A learning algorithm for ordered FDT induction has been presented and discussed in [7].

3.3 Construction of Structure Function Based on Ordered FDT

According to [5], FDTs allow developing fuzzy decision rules or a decision table in decision support system. A decision table contains all possible values of input attributes and the corresponding values of the output attribute that is calculated using the FDT. Because there is a unique correlation between FDT's attributes and values of variables and structure function, the decision table can be described in term of structure function (reliability analysis). In this case the inducted FDT allows forming a table that agrees with the structure function. This implies that all possible combinations of values of the component states (all state vectors) have to be analyzed by the FDT to classify state vectors into M classes of the system performance levels.

Each non-leaf node is associated with an attribute $A_i \in A$, or in terms of reliability analysis: each non-leaf node is associated with a component. The non-leaf node agreeing with attribute A_i has m_i outgoing branches. The s-th outgoing branch ($s = 0, \ldots, m_i - 1$) from the non-leaf node corresponding to attribute A_i agrees with state s of the i-th component ($x_i = s$). A path from the root to a leaf defines one or more state vectors (according to the values of the input attributes (component states) occurred in the path) for which the structure function takes value determined by the value of the output attribute. If any input attribute is absent in the path, all possible states have to be considered for the associated component.

The constructed structure function can be used for evaluation of MSS reliability through calculation of special measures and indices, in particular, the probability of system performance level (3).

4 Illustration of the Structure Function Construction

Let us consider the structure function construction based on uncertain data for a real system. As an example let us provide the analysis of the offshore electrical power generation system presented in [2]. The structure function for the evaluation of this system was constructed based on non-ordered FDT in [6]. In this paper, the structure function is constructed by the application of ordered FDT and the results will be compared.

4.1 The Construction of the Structure Function of the Offshore Electrical Power Generation System

The purpose of this system (Fig. 3) is to supply two nearby oilrigs with electric power. The system includes 3 generators: two main generators G_1 and G_3, and standby generator G_2. Both main generators are at oilrigs. In addition, oilrig 1 has generator G_2 that is switched into the network in case of outage of G_1 or G_3. The control unit U continuously supervises the supply from each of the generators with automatic control of the switches. If, for instance, the supply from A_3 to oilrig 2 is not sufficient, whereas the supply from A_1 to oilrig 1 is sufficient, U can activate A_2 to supply oilrig 2 with electric power through the standby subsea cables L. This implies that the system consists of 5 relevant components ($n = 5$): generators G_1, G_2, and G_3, control unit U, and the standby subsea cables L. Furthermore, according to the description of the system activity in [2], we assume that the system and all its components have 3 states/performance levels ($M = 3$ and $m_i = 3$, for $i = 1,2,\ldots,5$). Next, let us denote variables defining states of the system components in the following way: main generators G_1 and G_3 as x_1 and x_3 respectively, standby generator G_2 as x_2, and control unit U and standby subsea cables L as x_4 and x_5 respectively.

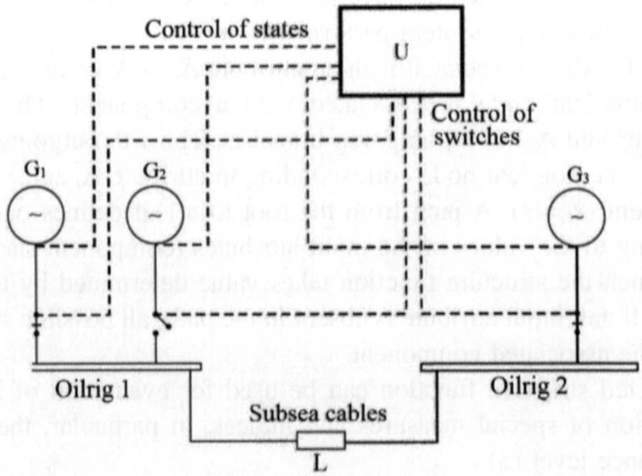

Fig. 3. Outline of the offshore electrical power generation system [2]

In [6], the repository of the offshore power generation system was formed. This repository collects 108 (from 243 possible) samples of the system behavior. The monitoring of this system permitted obtaining information about some combinations of component states and the corresponding performance levels of the system. However, this information is not complete. This uncertainty is caused by the ambiguity of classification of component states and system performance levels into classes of exact values [5, 10]. This data can be interpreted as quasi-fuzzy data that describes occurrence of every value of every attribute with some possibility ranging from 0 to 1. Traditional mathematical approach for system reliability analysis based on the structure function cannot be used in this case. The new method for construction of the structure function based on an FDT is used. This method allows reducing indeterminate values and obtaining a completely specified structure function.

The data obtained based on the monitoring of the offshore electrical power generation system is presented in Table 1 in form of repository. This table includes 6 columns. The first five columns agree with input attributes (or components states).The last column includes the output attribute values (or system performance level). Every input attribute has 3 values defined by number of states of the component. Similarly, the output attribute has 3 values too.

Table 1. A training set for the FDT induction

No	A_1			A_2			A_3			A_4			A_5			B		
	$A_{1,0}$	$A_{1,1}$	$A_{1,2}$	$A_{2,0}$	$A_{2,1}$	$A_{2,2}$	$A_{3,0}$	$A_{3,1}$	$A_{3,2}$	$A_{4,0}$	$A_{4,1}$	$A_{4,2}$	$A_{5,0}$	$A_{5,1}$	$A_{5,2}$	B_0	B_1	B_2
1	0.8	0.2	0.0	0.8	0.1	0.1	0.7	0.2	0.1	0.8	0.2	0.0	0.7	0.3	0.0	**0.7**	**0.3**	**0.0**
2	0.8	0.1	0.1	0.7	0.1	0.2	0.6	0.2	0.2	0.8	0.2	0.0	0.0	1.0	0.0	**0.8**	**0.1**	**0.1**
3	1.0	0.0	0.0	0.7	0.3	0.0	0.9	0.1	0.0	0.0	0.9	0.1	0.7	0.2	0.1	**1.0**	**0.0**	**0.0**
4	0.8	0.1	0.1	0.8	0.1	0.1	0.0	0.9	0.1	0.1	0.9	0.0	0.8	0.1	0.1	**0.2**	**0.7**	**0.1**
5	0.7	0.2	0.1	1.0	0.0	0.0	0.2	0.7	0.1	0.1	0.6	0.3	0.1	0.9	0.0	**0.0**	**0.7**	**0.3**
6	1.0	0.0	0.0	0.0	1.0	0.0	1.0	0.0	0.0	0.7	0.2	0.1	0.5	0.3	0.2	**0.7**	**0.2**	**0.1**
7	0.8	0.1	0.1	0.2	0.6	0.2	0.8	0.1	0.1	0.0	0.0	1.0	0.8	0.1	0.1	**0.6**	**0.2**	**0.2**
8	1.0	0.0	0.0	0.0	0.9	0.1	0.0	0.9	0.1	0.0	0.9	0.1	0.7	0.3	0.0	**0.0**	**0.6**	**0.4**
9	0.8	0.1	0.1	0.0	0.9	0.1	0.1	0.8	0.1	0.0	0.9	0.1	0.2	0.7	0.1	**0.1**	**0.6**	**0.3**
10	0.7	0.3	0.0	0.1	0.8	0.1	0.0	0.1	0.9	0.8	0.1	0.1	0.8	0.2	0.0	**0.7**	**0.1**	**0.2**
11	0.7	0.2	0.1	0.0	0.9	0.1	0.0	0.1	0.9	0.0	0.1	0.9	0.7	0.2	0.1	**0.0**	**1.0**	**0.0**
12	0.1	0.6	0.3	0.8	0.2	0.0	0.2	0.8	0.0	0.1	0.7	0.3	0.0	1.0	0.0	**0.1**	**0.4**	**0.5**
13	0.2	0.8	0.0	0.7	0.3	0.0	0.1	0.1	0.8	0.1	0.6	0.3	1.0	0.0	0.0	**0.0**	**0.7**	**0.3**
14	0.1	0.8	0.1	0.2	0.7	0.1	0.0	1.0	0.0	0.3	0.6	0.4	0.9	0.1	0.0	**0.2**	**0.8**	**0.0**
15	0.0	0.2	0.8	0.9	0.1	0.0	0.2	0.8	0.0	0.0	0.1	0.9	0.0	0.1	0.9	**0.0**	**0.6**	**0.4**
16	0.0	0.1	0.9	1.0	0.0	0.0	0.0	0.1	0.9	0.1	0.6	0.3	0.0	0.2	0.8	**0.2**	**0.5**	**0.3**
17	0.0	0.2	0.8	0.1	0.6	0.3	0.2	0.5	0.3	0.2	0.7	0.1	0.0	0.3	0.7	**0.1**	**0.1**	**0.8**
18	0.0	0.2	0.8	0.2	0.7	0.1	0.0	0.2	0.8	1.0	0.0	0.0	1.0	0.0	0.0	**0.7**	**0.0**	**0.3**
19	0.0	0.1	0.9	0.0	0.2	0.8	0.2	0.7	0.1	0.0	0.1	0.9	0.2	0.7	0.1	**0.0**	**0.1**	**1.9**
	…	…	…	…	…	…	…	…	…	…	…	…	…	…	…	**…**	**…**	
108	0.0	0.0	1.0	0.0	0.1	0.9	0.0	0.1	0.9	0.0	0.1	0.9	0.0	0.8	0.2	**0.0**	**0.0**	**1.0**

For example, the first row in Table 1 indicates the nonworking ($x_1 = 0$) and insufficient ($x_1 = 1$) states of generator A_1 with possibility of 0.8 and 0.2 respectively, while the possibility of the working state ($x_1 = 2$) is 0. In case of stable generator A_2, the state is indicated as nonworking ($x_2 = 0$) with possibility of 0.8 and as other values ($x_2 = 1$ and $x_2 = 2$) with possibilities of 0.1. State of main generator A_3 is nonworking ($x_3 = 0$) with possibility 0.7 and insufficient ($x_3 = 1$) or working ($x_3 = 2$) with possibilities 0.2 and 0.1 respectively. States of control unit U are defined as $x_4 = 0$ with possibility 0.8, $x_4 = 1$ with possibility 0.2 and $x_4 = 2$ with possibility 0. Only 2 of 3 states of the standby subsea cables L are relevant in this case because possibility of state $x_5 = 2$ is 0. The relevant states have possibilities 0.7 for $x_5 = 0$ and 0.3 for $x_5 = 1$. The system state is interpreted as a failure for this components states with the possibility 0.7 ($\phi(\boldsymbol{x}) = 0$) and as the sufficient state ($\phi(\boldsymbol{x}) = 1$) with the possibility 0.3, while the state of perfect operation ($\phi(\boldsymbol{x}) = 2$) is not indicated since its possibility is 0.

The repository in Table 1 includes input attributes $\mathbf{A} = \{A_1, A_2, A_3, A_4, A_5\}$ and the output attribute B for the offshore electrical power generation system in Fig. 3. Each input attribute is defined as: $A_i = \{A_{i,0}, A_{i,1}, A_{i,2}\}$, for $i = 1,2,\ldots,5$, and the output attribute is $B = \{B_0, B_1, B_2\}$. The values of the input attributes and the output attribute are defined in Table 1 and are used for the ordered FDT construction as the training test. The induction of the ordered FDT based on this training test can be implemented using the cumulative information estimates (4). This ordered FDT is presented in Fig. 4. The nodes of this FDT agree with the input attributes. Every node has 3 branches according to the values of the corresponding input attribute from the training test (Table 1). Every branch correlates with some values of the output attribute. The set of the output attribute values in a branch is denoted as a leaf if the analysis finishes and one of the values of the output attribute can be chosen according to algorithms proposed in [7, 8].

This ordered FDT (Fig. 4) can be used for the analysis of all possible states of system components to construct the structure function of the ffshore electrical power generation system. All possible component states (all state vectors) have to be used for calculation of the system performance level by the ordered FDT to form the decision table (structure function). Let us explain this idea in more detail.

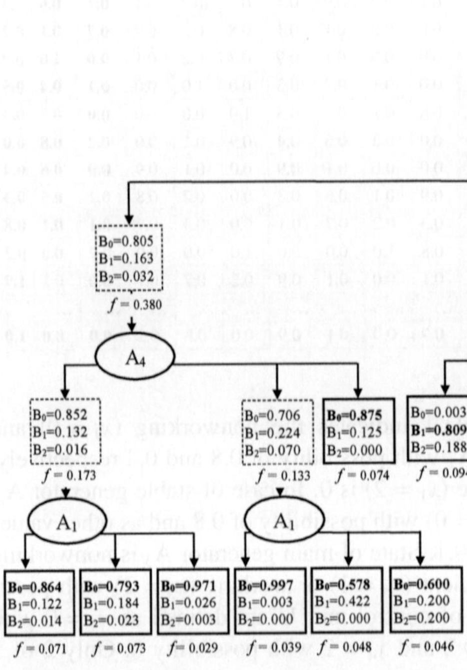

Fig. 4. Ordered FDT constructed based on the data obtained by the monitoring of the offshore electrical power generation system from Fig. 3

Preliminary analysis of the data obtained based on the monitoring (see Table 1) shows that possible values of the output attribute B are distributed as follows: value 0 – with confidence 0.493, value 1 – with confidence 0.209 and value 2 – with confidence 0.298. These values are implied by frequency of every output value in the training set (Table 1). Attribute A_3 is associated with the FDT root. So, analysis of the data starts from this attribute. This attribute has the following possible values: $A_{3,0}$, $A_{3,1}$, and $A_{3,2}$. Value $A_{3,0}$ of this attribute makes the output attribute B to be B_0 (the system is non-operational) with the confidence of 0.805. Other variants, B_1 and B_2, of output attribute B can be chosen with the confidence of 0.163 and 0.032 respectively. If the attribute A_3 has other values, i.e. $A_{3,1}$ or $A_{3,2}$, then the analysis is done similarly.

If the value of attribute A_3 is $A_{3,0}$, than the next attribute in the analysis is A_4, which has values $A_{4,0}$, $A_{4,1}$, and $A_{4,2}$. Value $A_{4,2}$ of this attribute agrees with a leaf representing the output attribute. Therefore, in this situation, the analysis is stopped and value of the output attribute is defined: value B_0 of attribute B should be chosen with the confidence of 0.875, and values B_1 and B_2 with confidences 0.125 and 0.000 respectively. Similarly, the process of the analysis of the ordered FDT continues for the other input attributes and their values.

Next, let us consider state vector $\boldsymbol{x}$ = (0, 0, 0, 0, 0). The analysis based on the FDT starts with the attribute A_3 (Fig. 4) that is associated with the 3-rd component. State of this component is 0 ($x_3 = 0$) for the specified state vector. Therefore, the branch for value $A_{3,0}$ of attribute A_3 value is taken into account. According to this value, the identification of the output attribute value (system performance level) has to continue through attribute A_4. According to the state vector, $x_4 = 0$, therefore, attribute A_4 has value $A_{4,0}$. Similarly, we continue this process. According to the state vector, $x_1 = 0$, therefore, attribute A_1 has value $A_{1,0}$. Now, value of the output attribute can be indicated because the branch has a leaf. So, value of the output attribute is defined as 0 with the confidence of 0.864. We can take this decision without analysis of other attributes (A_2 and A_5). Analysis of other state vectors is similar and allows obtaining all possible values of the system performance level in the form of the structure function. The analysis of all possible state vectors from $\boldsymbol{x}$ = (0, 0, 0, 0, 0) to $\boldsymbol{x}$ = (2, 2, 2, 2, 2) allows us to construct the complete structure function of the offshore electrical power generation system depicted in Fig. 4.

It is important to note that this method of construction of the structure function based on non-ordered or ordered FDTs permits to compute (restore) data missing from the monitoring.

4.2 Efficiency and Accuracy Investigation of the Proposed Method

In this section, we present a simple case study carried out in order to verify the modelling approach described in previous sections. A representation of the system using the structure function allows calculating different indices and measures for estimation of system reliability. The probabilities of system performance levels (3) are one of them. Suppose that the probabilities of the components states of the offshore electrical power generation system have values shown in Table 2 (according to data in [6]). In this case, the probabilities of the system performance levels are: $A_2 = 0.45$, $A_1 = 0.4$, and $A_0 = 0.15$.

Note, the offshore electrical power generation system has been analyzed in [6] based on equal data with application of non-ordered FDT. The probabilities of this system performance levels were calculated too: $A_2 = 0.52$, $A_1 = 0.31$ and $A_0 = 0.17$. There are differences between values of the system performance levels calculated based on different types of FDT. Therefore, the verification of the proposed method and its analysis has to be performed.

The verification includes two parts. The first is investigation of the accuracy of the proposed method for construction of the structure with application of ordered FDT. The second part of the verification is the comparison of accuracy of the results (structure functions) that are computed based on non-ordered and ordered FDTs.

We use structure function of the offshore electrical power generation system described in [2] to examine efficiency and accuracy of the proposed method for construction of the structure function based on uncertain data. This function in [2] is defined by the mathematical equation and can be calculated for all possible values of the system components states (variables of the function). In addition, the values of the structure function and its variables are exact integer values. Therefore, all integer values of components states and performance levels of the structure function in [2] have to be transformed to values with some possibilities. We can use algorithms from [21, 22] to transform data from numeric to fuzzy values in this case. In addition, the initial data in the proposed method is incompletely specified. This incompleteness is modeled by random deleting of some state vectors and the corresponding values of system performance levels. The range of deleted states was changed from 5% to 90%. Each transformed structure function can be interpreted as uncertain data obtained by the aforementioned monitoring. This data is formed as repository according to request presented in Sect. 3.1. We used this data to construct the structure function based on the proposed methods with application of the ordered FDT induction (Sect. 3.2). The structure function construction was implemented according to the concept introduced in Sect. 3.3.

Table 2. Components states probabilities

Component state, s	Probabilities				
	$p_{1,s}$	$p_{2,s}$	$p_{3,s}$	$p_{4,s}$	$p_{5,s}$
0	0.1	0.2	0.1	0.1	0.1
1	0.4	0.4	0.4	0.3	0.1
2	0.5	0.4	0.5	0.6	0.8

As a result, a single or a small group of state vectors might be misclassified. Therefore, we had to estimate this misclassification by the error rate. The constructed structure functions were compared with the exact specified function (it was defined at the beginning of the experiments), and the error rate was calculated as a ratio of wrong values of the structure function to the dimension of unspecified part of the function.

The experiments were iterated 1000 times for every system and fixed values of parameters α, β and numbers of unspecified state vectors. The effect of parameters α and β on the error rate illustrates graphs in Fig. 5. In addition, we provide the structure function reconstruction by the application of value that appears most frequently in the immediate neighborhood [23]. This method does not use specific analysis of the structure function and the error rate can be considered as maximal. The error rate of this method is denoted by red line and as "Maximal error" in graph (Fig. 5). The best decision (with minimal error rate) is in blue color.

The best result has minimal error rate. The results for investigated systems are shown in Table 3 in the second column, and the values for this result are $\alpha = 0.15$ and $\beta = 0.95$. The error rate depends on unspecified part of the initial data. This error

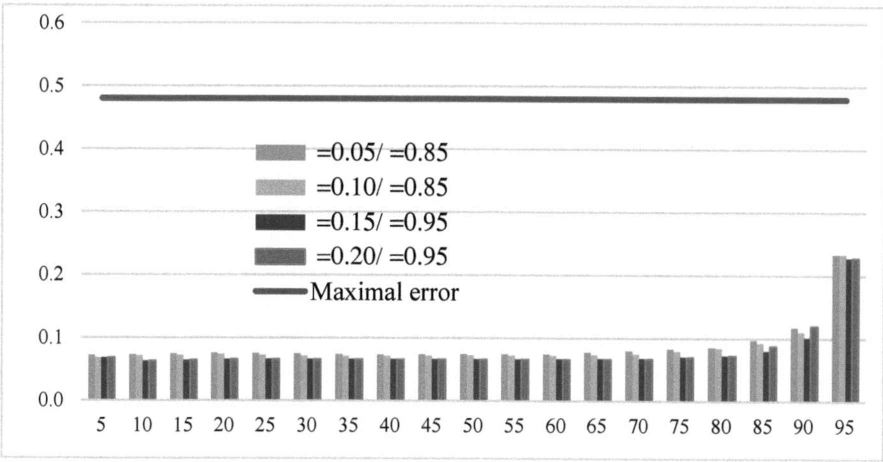

Fig. 5. The error rate for the construction of the structure function for the offshore electrical power generation system in [2] depending of different values of parameters α and β (Color figure online)

increases significantly if the unspecified part is more than 80% for all investigated systems. And we can see insignificant growth of the error rate if the unspecified part is less than 10%. The values of parameters α and β have been defined for the best decision based on these experiments (Table 3).

Therefore, analysis of the error rate for the proposed method for the construction of the structure function based on ordered FDT shows that this method has good efficiency. This method is acceptable for the incomplete data and the incompleteness of initial data can be indicated from 10% to 85%. The constructed structure function by the proposed method has less error rate than maximal error rate in interval of the incompleteness.

Let us to consider the comparison of the proposed method of the structure function construction based on uncertain data with application ordered FDT with the similar method in which the non-ordered FDT is used instead of ordered FDT. The method for structure function construction has been introduced in [5, 6]. The offshore electrical power generation system has been analyzed by this method in [6]. The accuracy of the method with application of non-ordered FDT has been estimated similarly in [6]. The result of this estimation is shown in the last column in Table 3. The comparison of the accuracies of two methods in Table 3 shows that there are insignificantly differences of accuracies of these methods. The new method for the structure function construction based on uncertain data with application of ordered FDT has higher accuracy.

The differences of the accuracies of the methods with application of non-ordered FDT and ordered FDT cause that the structure functions constructed by these methods are not same. Therefore, as was mentioned above, the values of the probabilities of the system performance levels are different for these methods. The accuracy evaluation for the considered system reveals that the ordered FDT has the higher accuracy, therefore,

Table 3. The error rate for the construction of the structure function of the offshore electrical power generation system

Unspecified state vectors, in %	The error for the best result calculated based on ordered FDT ($\alpha = 0.15/\beta = 0.95$)	The error for the best result calculated based on non-ordered FDT ($\alpha = 0.15/\beta = 0.85$)
5	0.0681	0.0661
10	0.0627	0.0637
15	0.0643	0.0682
20	0.0658	0.0661
25	0.0659	0.0670
30	0.0660	0.0662
35	0.0661	0.0648
40	0.0661	0.0663
45	0.0661	0.0657
50	0.0661	0.0659
55	0.0664	0.0671
60	0.0666	0.0673
65	0.0674	0.0679
70	0.0681	0.0700
75	0.0703	0.0743
80	0.0725	0.0813
85	0.0796	0.0995
90	0.1014	0.1465
95	0.2319	—

the performance levels probabilities of the offshore electrical power generation system can be indicated by the structure function constructed with application of the ordered FDT ($A_2 = 0.45$, $A_1 = 0.4$ and $A_0 = 0.15$).

5 Conclusion

The problem of the development of mathematical model based on uncertain data of investigated system is current in reliability engineering [24–26]. The reviews of this problem in [24, 26] systematise three diverse approaches under uncertainty in terms of their representation of uncertain quantities and analytical output. The authors distinguish three methodologies (semi-quantitative uncertainty factors, probability bounds analysis, and Robust Decision Making) that decided the problem of uncertainty in very different ways. Other methodology for analysis of human reliability taking into account the uncertainty of initial data is considered in [25]. The important aspect of this methodology is application of fuzzy data. According to these methodologies new mathematical models are introduced in reliability engineering. This results in the development of new algorithms for the reliability indices and measures calculation. At the same time, there are many effective methods and algorithms for the evaluation of system reliability based on typical mathematical model, such as structure function [1, 2, 4], decision diagram [27], reliability block diagram [1], fault tree [2], etc.

This paper proposes the new method to construct a structure function based on uncertain data by application of methodology of ordered FDT and decision table. The structure function is one of often used mathematical models in reliability engineering, and there are a lot of algorithms to calculate different reliability indices and measures based on the structure function [1, 2, 28]. The presented method represents development of the approach considered in [5, 6] where the application of Data Mining methods for reliability analysis have been introduced for uncertain and incomplete initial data. In these papers the non-ordered FDT has been used for the structure function construction. It is a simple type of FDT that has no special properties. In this paper, another type of FDT, ordered FDT, was used. The ordered FDT has the same attributes in each level of the tree. It means that next attribute can be analysed (calculated or measured) independently of the result of analysis at the current level. Such structure allows using parallel procedure in analyses of some attributes. But for complex system the ordered FDT can have more complex structure (has more nodes and/or levels) according to [7, 8]. At the same time, the ordered FDT can have higher accuracy according to the conducted experiments. The accuracy of the method for examined system is acceptable. Future work will be devoted to investigation of more complicated cases.

The correlation of the system dimension and the method efficiency (time required and error) will be considered in the future too. Now, we can use the FDT application to forecast the dimension and specificity of the structure function. The construction of the structure function of 50 state vectors is possible based on 20–30 state vectors (40–60% of defined state vectors). The structure function construction based on 5–10 state vectors (10–20%) is possible too. But the level of accuracy depends on the quality of this set of state vectors. It will be essential to continue the verification and validation of our proposed method with data sets of different properties and sizes.

Acknowledgment. This work is supported by the grant of Ministry of Education, Science, Research and Sport of the Slovak Republic VEGA 1/0038/16.

References

1. Lisnianski, A., Levitin, G.: Multi-state System Reliability. Assessment, Optimization and Applications. World Scientific, Singapore (2003)
2. Natvig, B.: Multistate Systems Reliability Theory with Applications. Wiley, New York (2011). doi:10.1002/9780470977088
3. Aven, T.: Reliability evaluation of multistate systems with multistate components. IEEE Trans. Reliab. **R-34**, 473–479 (1985). doi:10.1109/TR.1985.5222235
4. Kvassay, M., Zaitseva, E., Levashenko, V.: Minimal cut sets and direct partial logic derivatives in reliability analysis. In: Proceedings of the European Safety and Reliability Conference on Safety and Reliability: Methodology and Applications, pp. 241–248. CRC press, (2015). doi:10.1201/b17399-37
5. Zaitseva, E., Levashenko, V.: Construction of a reliability structure function based on uncertain data. IEEE Trans. Reliab. **65**, 1710–1723 (2016). doi:10.1109/TR.2016.2578948

6. Zaitseva, E., Levashenko, V., Kvassay, M.: Application of structure function in system reliability analysis based on uncertain data. In: 12th International Conference on ICT in Education, Research and Industrial Applications: Integration, Harmonization and Knowledge Transfer, pp. 361–374, CEUR-WS (2016)
7. Levashenko, V., Zaitseva, E., Puuronen, S.: Fuzzy classified based on fuzzy decision tree. In: IEEE International Conference on Computer as a Tool (EUROCON 2007), pp. 823–827. IEEE Press (2007). doi:10.1109/EURCON.2007.4400614
8. Levashenko, V.G., Zaitseva, E.N.: Usage of new information estimations for induction of fuzzy decision trees. In: Yin, H., Allinson, N., Freeman, R., Keane, J., Hubbard, S. (eds.) IDEAL 2002. LNCS, vol. 2412, pp. 493–499. Springer, Heidelberg (2002). doi:10.1007/3-540-45675-9_74
9. Kuo, W., Zhu, X.: Importance Measures in Reliability, Risk, and Optimization: Principles and Applications. Wiley, Chichester (2012). doi:10.1002/9781118314593
10. Aven, T., Zio, E., Baraldi, P., Flage, R.: Uncertainty in Risk Assessment: The Representation and Treatment of Uncertainties by Probabilistic and Non-Probabilistic Methods. Wiley (2014). doi:10.1002/9781118763032
11. Cutello, V., Montero, J., Yanez, J.: Structure functions with fuzzy states. Fuzzy Sets Syst. **83**, 189–202 (1996)
12. Stanton, N.A., Baber, C.: Error by design: methods for predicting device usability. Des. Stud. **23**, 363–384 (2002). doi:10.1016/S0142-694X(01)00032-1
13. Huang, H.-Z.: Structural reliability analysis using fuzzy sets theory. Eksploatacja i Niezawodnosc Maintenance Reliab. **14**, 284–294 (2012)
14. Witten, I.H., Frank, E., Hall, M.A.: Data Mining: Practical Machine Learning Tools and Techniques. Elsevier, Burlington (2011)
15. Mitra, S., Konwar, K.M., Pal, S.K.: Fuzzy decision tree, linguistic rules and fuzzy knowledge-based network: generation and evaluation. IEEE Trans. Syst. Man Cybern. Part C Appl. Rev. **32**, 328–339 (2002). doi:10.1109/TSMCC.2002.806060
16. Tsipouras, M.G., Exarchos, T.P., Fotiadis, D.I.: A methodology for automated fuzzy model generation. Fuzzy Sets Syst. **159**, 3201–3220 (2008). doi:10.1016/j.fss.2008.04.004
17. Quinlan, J.R.: Simplifying decision trees. Int. J. Man Mach. Stud. **27**, 221–234 (1987). doi:10.1023/A:1022643204877
18. Kruse, R., Doring, Ch., Lesot, M.-J.: Fundamentals of fuzzy clustering. In: Valente de Oliveira, J., Pedrycz, W. (eds.) Advances in Fuzzy Clustering and its Applications. Wiley (2007). doi:10.1002/9780470061190.ch1
19. Tanaka, H., Fan, L.T., Lai, F.S., Toguchi, K.: Fault-tree analysis by fuzzy probability. IEEE Trans. Reliab. **R-32**, 453–457 (1983). doi:10.1109/TR.1983.5221727
20. Yuan, Y., Shaw, M.J.: Induction of fuzzy decision trees. Fuzzy Sets Syst. **69**, 125–139 (1995). doi:10.1016/0165-0114(94)00229-Z
21. Lee, H.M., Chen, C.-M., et al.: An efficient fuzzy classifier with feature selection based on fuzzy entropy. IEEE Trans. Syst. Man Cyber Part B Cybern. **31**, 426–432 (2001). doi:10.1109/3477.931536
22. Shyrai, S., Viattchenin, D.: Clustering the intuitionistic fuzzy data: detection of an unknown number of intuitionistic fuzzy clusters in the allotment. In: International Conference on Information and Digital Technologies (IDT 2015), pp. 314–323. IEEE Press (2015). doi:10.1109/DT.2015.7222990
23. Braga-Neto, U.M., Dougherty, E.R.: Exact correlation between actual and estimated errors in discrete classification. Pattern Recogn. Lett. **31**, 407–412 (2010). doi:10.1016/j.patrec.2009.10.017

24. Shortridge, J., Aven, T., Guikema, S.: Risk assessment under deep uncertainty: a methodological comparison. Reliab. Eng. Syst. Saf. **159**, 12–23 (2017). doi:10.1016/j.ress.2016.10.017
25. Baraldi, P., Podofillini, L., et al.: Comparing the treatment of uncertainty in Bayesian networks and fuzzy expert systems used for a human reliability analysis application. Reliab. Eng. Syst. Saf. **138**, 176–193 (2015). doi:10.1016/j.ress.2015.01.016
26. Dubois, D.: Representation, propagation, and decision issues in risk analysis under incomplete probabilistic information. Risk Anal. **30**, 361–368 (2010). doi:10.1111/j.1539-6924.2010.01359.x
27. Zaitseva, E., Levashenko, V., Kostolny, J., Kvassay, M.: A multi-valued decision diagram for estimation of multi-state system. In: IEEE EuroCon 2013, pp. 645–650. IEEE Press (2013). doi:10.1109/EUROCON.2013.6625049
28. Zaitseva, E., Levashenko, V., Kostolny, J.: Multi-state system importance analysis based on direct partial logic derivative. In: 2012 International Conference on Quality, Reliability, Risk, Maintenance, and Safety Engineering, pp. 1514–1519. IEEE Press (2012). doi:10.1109/ICQR2MSE.2012.6246513

Using Choreography of Actors and Rewriting Rules to Adapt Legacy Fortran Programs to Cloud Computing

Eugene Tulika(✉), Anatoliy Doroshenko, and Kostiantyn Zhereb

Institute of Software Systems of National Academy of Sciences of Ukraine, Glushkov prosp. 40, Kiev, 03187, Ukraine
eugene.tulika@gmail.com, doroshenkoanatoliy2@gmail.com, zhereb@gmail.com

Abstract. We propose an approach to the semi-automatic transformation of legacy Fortran applications for execution on cloud computing platforms. The architecture is based on the choreography concept implemented as actor model, which allows to improve the scalability of the system and reduces overhead on message passing. The rewriting rules technique is used to formally define certain transformation steps. The approach is tested on an example program from the quantum chemistry field. The main contribution of this chapter is (1) using rewriting rules technique to automate transformation steps and (2) using service choreography to optimize resource allocation on the cloud.

Keywords: Cloud computing · Scalable parallelism · Actor model · Rewriting rules · Choreography

1 Introduction

Fortran language exists since the 1950s and during this time positioned itself as the best tool for scientific research. The language has huge support from the software industry, new libraries and compilers are released that allow using modern technologies and standards. Examples are: compiler from Portland Group [1] for Fortran supporting GPGPU and CUDA; High-Performance Fortran Forum [2] creates compilers and standards for high-performance Fortran; OpenMP and MPI for parallel and distributed computation allow to use Fortran for modern cluster computing; Intel popularizes Fortran [3] in order to support proprietary Intel compilers and programs; Coarray Fortran language fork became a standard in Fortran 2003 and allows to support computation on distributed arrays [4]; there are many Fortran libraries [5] for numerical analysis. Besides, conservative policy for backward compatibility makes code written on old standards compatible with the new compilers. All this makes Fortran an attractive platform for scientific research. However, Fortran suffers from outdated standards that make it hard to write an efficient code for computations with distributed memory. Another problem is a great amount of legacy code which was written without taking into account distributed architectures.

Cloud computing became popular because of the need to reduce the cost of computations. The difference between cloud computing and other parallel approaches is usage of

A. Ginige et al. (Eds.): ICTERI 2016, CCIS 783, pp. 76–96, 2017.
https://doi.org/10.1007/978-3-319-69965-3_5

scalability (ability to support more resources) vs. performance (running fast on given resources). Scalable systems can leverage computing instances which individually perform poorly. However, using a large number of such instances and making them work together allows building systems which outperform the most powerful mainframes. Cloud computing is focused on reducing the cost and time/performance balance using cheap components, and therefore allows using more resources. Therefore, there is a need for methodologies of writing the software which can leverage multiple independent low-performant instances to achieve same results as on the single supercomputer.

This chapter describes the approach of adapting legacy Fortran scientific applications for the cloud platforms. We propose a methodology for the semi-automatic transformation of the legacy application to the scalable programs leveraging multiple computing instances to produce better performance. The main contribution of this chapter is (1) using rewriting rules technique to automate transformation steps and (2) using service choreography to optimize resource allocation on the cloud.

The generic question of Fortran parallelization for distributed architecture is widely studied [14, 15]. This chapter continues the research on parallelizing legacy Fortran code that was started in [31]. While the paper [31] focused on shared-memory multicore platform, this chapter targets a more complex distributed cloud platform. The preliminary results of choosing parallelization model for cloud platform were presented in [32]; this chapter elaborates on the previous results by applying the choreography to the actor model in order to express distributed processes. We propose a solution to the transformation of the legacy code into the choreographed actor model. It provides a robust framework for the development of applications that can be distributed across cloud infrastructure without additional efforts, while usage of MPI for the highly scalable applications requires manual control over data placement and inter-process communication.

2 Scalability Property of the System

Distributed Cloud systems are characterized by two types of performance challenges they deal with:

1. The pressure of input information – occurs in signal system and with any kind of high volume streams of data.
2. Computation of the high volume of the static data – occurs in situations where information is already captured and does not change but the size of the data is big relatively to the amount of the computations which need to be performed. It happens in a variety of scientific problems.

The scalable application has different strategies to deal with both problems. If the stream of input information grows, cloud infrastructure should be able to add more operational resources to capture the information and persist it in the data storage. Different strategies can be used, such as using more network bandwidth, faster drives for the storage, more resources which receive data, etc. Dealing with data streams of variable size requires the concept of auto-scaling – a capability to adaptively add computing resources when certain metrics go up, i.e. memory pressure or queues size.

For the case when computation should be performed on large sets of data, cloud infrastructure should be able to add more computing instances to address the bigger problem. Computing instances can be any machines – either on bare metal or virtualized with the processor and RAM. An approximate number of resources needed to calculate the problem with the data of certain size can be predicted based on empirical data of the previous program executions, as well as time and space complexity of the program. The dependency between the size of the problem, a number of available resources and a time to compute the program can be expressed as a function *time* = *f(input_size, resources_amount)* and characterizes the scalability of the application. If input size is fixed, the number of resources is growing, but the function is not monotonously decreasing, it means that application cannot leverage new resources to calculate the bigger problem. Scale factor is a ratio z between two functions $t_1 = f_1(s_1, r_1)$ and $t_2 = f_2(s_2, r_2)$ where s_1, s_2 are input sizes; r_1, r_2 are numbers of computing instances; t_1, t_2 are execution times of the distributed program. The distributed program is perfectly scalable if the following is true:

$$r_1/r_2 = t_1/t_2 \text{ when } s_1 = s_2 \tag{1}$$

The program is un-scalable if the following is true:

$$\lim_{r_1/r_2 \to \infty} \left(\frac{t_1}{t_2} \right) = 1 \tag{2}$$

This formula expresses the main characteristic of the scalable system: for the growing input size we can also increase the number of computational resources so that the time of computation stays the same. Every program can be changed to improve scale factor. This chapter describes one of the approaches to transform the legacy program so that better scalability can be achieved. There are three main problems which should be addressed before the program can be transformed to the scalable architecture:

1. Program subroutines has too many data dependencies and cannot be separated.
2. The program is designed to run on a constant number of processors. It does not allow to leverage more computation resources.
3. Program algorithm is not designed to divide and conquer the input data.

3 Our Approach

In this section we discuss our approach of transforming legacy applications to port them to modern parallel architecture and improve their performance metrics (including scalability). We build formal models of legacy code and apply semi-automated transformations to these models.

Model manipulation is performed using rewriting rules technique, implemented by rewriting rules engine TermWare [9]. The data in Termware is represented as *terms*, i.e.

tree-like expressions of a form $f(t_1, \ldots t_n)$, where t_i are either terms or simple types (numbers, strings, booleans). Terms can be considered as a data format, similar to XML or JSON – any structured data can be represented as terms.

Transformations are described as Termware *rules*, i.e. expressions of form `source [condition]-> destination [action]`. Here `source` is a source term (a pattern that is matched to the whole input model), `condition` is an additional condition of rule application, `destination` is a transformed term, `action` is an additional action that is performed when the rule fires. Each of 4 components of a rule can contain variables (denoted as `$var`) so that the rules are more generally applicable. The components `condition` and `action` are optional. They can execute any procedural code implemented in Java class (known as FactsDB). In particular, it can use the additional data on the program, e.g. identifiers table. Examples of rewriting rules are discussed in Sect. 6.

Transformation of legacy application usually includes the following steps:

(1) Building low-level syntactic model of source code;
(2) Building high-level algebraic model of application;
(3) Selecting target platform for transformation;
(4) Transforming high-level model so that it is compatible with the target platform;
(5) Implementing infrastructure code needed to support application on the target platform;
(6) Generating source code for the transformed application;
(7) Executing the transformed application on the target platform and evaluating its performance metrics.

We attempt to automate as many transformation steps as possible. However, some steps currently are performed manually. Therefore, our approach can be described as *semi-automated transformation*.

Step 1 is performed using a parser for the language of legacy application (such as Fortran in this chapter). This step is fully automated; however, a parser supporting given language is needed. We use a parser based on GCC Fortran Compiler [10]. The result of this step is a low-level syntactic model, which is AST (abstract syntax tree) represented as TermWare term.

Step 2 is more complex. During this step, we try to abstract away syntax details specific to a given language. As a result, we want to create a high-level model representing algorithms used in the legacy application. In many cases, parts of this high-level model can be created in an automated way by finding special patterns in the low-level model. In such cases, we manually identify these patterns and create rewriting rules based on them. Such rules are then applied automatically to the entire model. The advantage of such approach manifests for repeating pattern. If some algorithm, such as matrix multiplication, is used multiple time throughout the legacy code, and if all instances are similar (except for parameters such as variable names and matrix sizes), then we can create a single rule that will recognize all such instances.

As an alternative, we can manually describe a high-level model of application or some parts of the application. This approach can be used when creating algorithms from scratch, rather than extracting them from existing code. Also, such approach can be simpler to implement even for legacy code – provided that a developer creating the model understands details of the algorithm. In this case, the algebraic toolkit can be used to edit the model in graphical user interface [30].

Step 3 is choosing the most suitable parallel platform to execute transformed application. Possible platforms include shared-memory multicore systems, distributed clusters, Grids, Cloud environments (as used in this chapter), specialized hardware such as GPUs or FPGAs, etc. This step also includes more detailed choices, such as selecting cloud provider or implementation libraries/frameworks. This step is driven both by performance requirements – which platform will allow executing the given application with the best metrics – and other considerations such as availability of various execution environments, their costs, etc. Currently, this step is performed manually. It is possible to consider an intelligent assistant that would provide recommendations about which platforms would provide the best performance metrics for a given algorithm. Such assistant would allow developers to make informed decisions about performance impacts of choosing different platforms. It could be implemented either using performance simulations or by generating sample applications for different platforms and directly measuring their performance. However, currently, such systems are beyond the scope of our research.

Step 4 adapts algorithm (represented as a high-level model) to the target platform. For example, we may need to explicitly identify different tasks, build data dependency graph between them, implement data transfers, etc. As with step 2, such transformations are currently semi-automated. We implement a number of rewriting rules that cover common cases. A developer has to choose which rules apply for a given application, and possibly mark the regions of code where they should apply (e.g. choose which loops to parallelize). The rules are then applied automatically across entire model. Also, a developer may need to implement the rules that were not implemented yet. Eventually, we plan to create a library of reusable rules that cover most of the common platforms. An example of performing step 4 for cloud platform is discussed in Sects. 4 and 6.

Step 5 is needed to provide some "helper" code (methods, classes, components, ...) used to run transformed code on the target platform. Typically, such code is written once for a given platform and then reused for multiple applications running on this platform. We discuss an example of performing step 5 in Sect. 9.

Step 6 requires a code generator for a target language that converts a high-level model into a source code. As with step 1, this step can be fully automated, although the code generator should be developed for each target language. In general, developing code generator is easier than developing parser (step 1) or transforming a low-level model into a higher-level model (step 2). This is because higher-level models contain more information about the structure of algorithms used in the application.

Finally, on step 7 the performance of generated code is evaluated. Currently, this is done manually, by running the application on given inputs and measuring execution time. Additional tools such as profilers can be used. An example of such evaluation is discussed in Sect. 10.

The transformation of legacy applications is an iterative process. After completing steps 1–7 and evaluating results, we may repeat some of these steps to improve the performance of the transformed application. For example, a developer may initially choose to parallelize a certain code fragment that was identified as a hotspot (i.e. it takes a long time to execute in sequential code). After parallelizing this part of the code and profiling the transformed application, another hotspot might be identified. In this case, the process is repeated to obtain a new version of transformed application, hopefully with better scalability and performance.

4 Legacy Fortran Program

The problem of data dependencies between components of the program needs to be investigated in the first place. To perform such investigation, we build a high-level algebraic model of the program. As in the earlier paper [10], TermWare [9] can be used to build high-level algebraic models of Fortran code by applying transformations to the low-level compiler output. For more complex applications, a model can be built or enhanced manually.

In this chapter, we discuss our approach of transforming legacy scientific applications based on an example Fortran application from the quantum chemistry field. Some properties we discuss in this section are specific to this application, such as time distribution of subroutines or the structure of control flow graph. However, our approach can be applied in the more general case.

The application which will be used as an example to apply proposed transformation approach is called QuPoly and calculates the geometry of electron orbitals [8]. The application was initially optimized for single-core performance, without any usage of parallelism.

The first step of parallelizing transformation requires identifying the most promising subroutines for parallelization. The example program consists of data input, computation subroutines, logging intermediate results into the file and data output. Profiling the application before optimization shows the time distribution for the main computational steps:

- Data input from files and initialization – approximately 1% of time;
- Subroutine `hcore` calculates integrals for every atom – 60% of time;
- Subroutine `iterc` – optimization of the geometry of molecule, 30% of the time.

Therefore, the main focus of parallelization is in the subroutines `hcore` and `iterc`. Time performance of algorithms in both subroutines has a linear dependency on the input size. Input data of the program is the file with the geometry of atoms. Each of two subroutines is executed independently on every atom in the input data. Provided that there is no data dependency between segments of the code executed for every atom, calculation of the atoms can be performed in parallel. However, overall steps of the algorithm are sequential. Subroutine `iterc` depends on the results of `hcore`, because the calculation of the Fock operator requires integrals computed in `hcore`. Here is an example of the high-level algebraic model for the QuPoly program:

```
[ForEach(atom, InputData, [hcore(atom)]),
 ForEach(atom, InputData, [iterc(atom)])]
```

Structurally, the application can be modeled with the following graph:

$$I \rightarrow A([a_{1..n}]) \rightarrow B([b_{1..n}]) \rightarrow O \tag{3}$$

Graph vertices I, A, B, O stand for the sequential steps of computation: I – data input and allocation of the memory, A – calculation of the integrals for every atom (hcore), B – calculation of the Fock operator for every atom (iterc), O – final steps and data output. Graph edges correspond to the control flow between parts of the program, $a_{1..n}$ and $b_{1..n}$ stand for input data of subroutines. The goal of this step is to transform program to the parallel form:

$$I \rightarrow A_{1..n}([a_{1..n}]) \rightarrow B_{1..n}([b_{1..n}]) \rightarrow O \tag{4}$$

We transform the sequential loop A, into n parallel processes $A_{1..n}([a_{1..n}])$. Each process A_i will perform calculations on a single segment of data a_i. The same transformation will be applied to B as well. After manual analysis of the data dependencies, it was identified that there are no dependencies between iterations of the loop in code fragment A, and processes $A_{1..n}$ can be invoked in parallel, same applicable to the processes $B_{1..n}$. However, B has a dependency on the results of A and to address this dependency, a barrier synchronization will be used between parallel processes A_i and B_i.

5 Application of Actor Model to the Legacy Programs

There are multiple reasons why software program cannot make use of more computing instances to run more efficiently, which is related to the second problem of scalable architecture. Usually, the reason is in the implementation of the program: (a) it is not possible to run subroutines of the program on separate processes; (b) internally, subroutines rely on the shared memory to exchange the data. In this chapter, we use Actor model [17] to analyze and address this problem of the scalable architecture. Actor Model is a model of concurrent computations where the concurrent process is designed using primitive – an actor. There are set of rules imposed on the actor. The actor is activated by the parent process. It starts execution upon receiving a message from another actor and for every cycle of execution it can:

1. Execute local operation changing its state;
2. Send a finite number of messages to other actors communicating its local state;
3. Create a finite number of new actors.

Actors communicate by asynchronous and reliable message passing, i.e., whenever a message is sent it must eventually be received by the target actor [18]. Send and receive operations on the message are non-blocking and can happen independently.

Formally, the program represented by actor model can be modeled with an asynchronous network, represented by n-node directed graph $G = (V, E)$. V is a set of nodes,

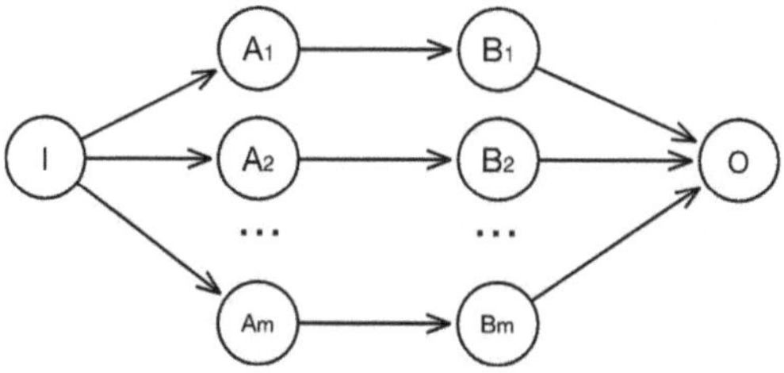

Fig. 1. Actor model applied to QuPoly program

where each node $v_i \in V$ corresponds to the location of the actor. E is a set of directed edges between nodes, where each directed edge $e_{ij} \in E$ corresponds to the message channel between v_i and v_j. T is a set of all possible event types: <send message>, <receive message>, <activate actor>, <execute operation>. S is a set of states of the process $<s_0..s_N, t_{>}$, where s_0 is the initial state, t is the final state. C is an execution plan of the program, defined as a set of events from initialization to termination. C is defined as a function $S \times E \rightarrow S$.

An important limitation is that actors are allowed to communicate with each other only through the message passing. Therefore, an application using the actor model cannot have the global state implemented via the shared memory. However, each actor can have its local state. It allows to relax the rules on the mapping of the actors to computing instances: any two actors can be allocated on different computing instances because they do not require to share the memory. For the purpose of this chapter we distinguish different types of the communication which happen between actors:

- Notification – these are tiny messages that should be delivered fast from a single actor to multiple other actors. Notifications are needed to communicate the local state of the actor to other actors through the message exchange. Other actors act upon notifications which affect the global state of the process. This type of messages has small to none overhead for the send or receive operations. However, an undelivered message can block the execution of the overall process, because the global state of the process cannot change without receiving notification from the actor.
- Data passing – this is the type of communication needed to exchange the results of data intensive operation between actors. Usually, the receiving actor that received a notification about completion of computation from another actor establishes a peer-to-peer connection to the sending actor that has the result of the operation. After that, the receiving actor downloads the data to the local storage. If the pair of actors is located on different remote computing instances, this operation can be more expensive than the computation itself.

Reasoning about application architecture in terms of Actor Model solves two issues: how to make independent sections of the program execute on separate computing instances and how to exchange data between them. When applying actor model to the architecture of the system, we implement subroutines of the large application as separate actors. The control flow graph (4), which was used to represent QuPoly program, is transformed into asynchronous network graph, represented on the Fig. 1.

Vertices of the graph correspond to the actors that implement subroutines of the QuPoly program and edges correspond to the channels between them. The execution of the program is performed according to the following steps. The input actor, I in the Fig. 1, is activated upon an event from the file system when the input data is allocated to the storage. It invokes Fortran code responsible for initial data processing. After the event is processed, the process I_1 sends broadcast message to all the channels IA_i with the notification about data been preprocessed. In the simplest case, this message is received by all the processes $A_{1..n}$, and they start execution invoking Fortran program. In the same way, the notification about completed calculation is sent to all the AB_{ij} channels.

There are different strategies for choosing the number of actors based on the input data size. In the example provided, if the number of the atoms in input data is n and the number of actors is m, then there are two simpler cases $n < m$ and $n = m$. In both cases, the data will be separated by the atoms and every actor will process a single atom. In the third case, when $n > m$, the actor I orchestrates which set of atoms should be sent to each A_i actor.

During the transformation of the legacy Fortran application to the Actor model, every independent segment of the program which is converted to an actor requires fulfillment of following requirements:

1. Coordination of work between actors should be described with the execution plan.
2. Segments should use message passing for communication.
3. Every segment should not have access to the global state and should not have side effects.

6 Automatic Code Transformation Using TermWare

The first requirement to transform legacy Fortran program to the actor model is to ensure that all its subroutines do not use global application state and code fragments which will be executed in parallel do not have any side effects. It can be done by replacing subroutines with pure functions (introduced in Fortran 95 standard). The following changes in code are needed:

- Create FUNCTION instead of SUBROUTINE
- Remove IMPLICIT statements – all local variables of the function should be explicitly declared.
- Remove COMMON BLOCKs (global variables) – all corresponding global variables should be passed as an inputs/outputs of the function, and reads/writes to global variables performed by the calling code.
- Remove read and write operations – all such operations should be invoked before and after the call to the pure function.

These changes can be implemented using TermWare rules [9]. As an example of transformation we use a set of rules to transform IMPLICIT statements to explicit declarations of variables:

```
1. _MarkPure(Subroutine($name,$params,$return,$body))
   -> Function($name,$params,$return,_MkImp($body))
2. _MkImp([$x:$y]) -> [_MkImp($x):_MkImp($y)]
3. _MkImp(NIL) -> NIL
4. _MkImp(Declare($var,$type,$val) ->
   Declare($var,$type,$val) [check($var, $type)]
5. [_MkImp(Assign($var,$expr)): $y] [isUnchecked
   ($var)] -> [Declare_MARK($var, $type) : [Assign
   ($var,$expr) : $y ]] [inferType ($var,$type)]
6. [$x:[Declare_MARK($var,$type) : $y]] ->
   [Declare_MARK($var,$type):[$x:$y]]
7. Function ($name,$params,$return,
   [Declare_ MARK($var,$type):$y]) -> Function
   ($name,$params,$return,[Declare($var,$type):$y])
```

Rule 1 triggers transformation, marking the body of the function with the marker term `_MkImp`. Rules 2 and 3 walk through the body of the function and expand the `_MkImp` marker to all operations. Rule 4 memorizes the variables which have explicit declaration using the method `check($var, $type)` from the FactsDB. Rule 5 finds variables without explicit declaration using method `isUnchecked($var)`. For these variables it determines the type with the method `inferType($var,$type)` and adds declaration marked with `Declare_MARK($var,$type)`. To determine the variable type, the method checks the variable name against IMPLICIT statements in Fortran code, as well as default convention that declares variables starting with `'i'-'n'` as INTEGER and all others as REAL. Rule 6 moves this declaration to the beginning of the function, and rule 7 removes the mark. As a result, term `Declare($var,$type)` is generated, which later is transformed to the declaration of the variable in the code.

Table 1. Initial and transformed source code for removing IMPLICIT declarations

Initial code	Transformed code
SUBROUTINE MATR_MULT(N,A,B,C)	*FUNCTION* MATR_MULT(N,A,B)
INTEGER,INTENT(IN) :: N	INTEGER,INTENT(IN) :: N
REAL*8,INTENT(IN)::A(N,N),B(N,N)	REAL*8,INTENT(IN)::A(N,N),B(N,N)
REAL*8,INTENT(OUT) :: C(N,N)	REAL*8 :: *MATR_MULT*(N,N)
	INTEGER :: I, J, K
	REAL*8 :: S
DO I=1,N	DO I=1,N
DO J=1,N	DO J=1,N
S = 0.0D+00	S = 0.0D+00
DO K=1,N	DO K=1,N
S=S+A(I,K)*B(K,J)	S=S+A(I,K)*B(K,J)
END DO	END DO
C(I,J)=S	*MATR_MULT*(I,J)=S
END DO	END DO
END DO	END DO

As an example of rules application, consider a simple procedure of square matrix multiplication (Table 1).

In the original code, some variables are used with no declaration. After transformation, all the variables are declared. Also, note that the syntax for SUBROUTINE is different from the FUNCTION. But such changes should not be described as additional rules. A simple substitution of the term "Subroutine" with the term "Function" is sufficient. During the code generation phase, all necessary changes are added automatically, which is one of the advantages of the TermWare and high-level algebraic models [10].

Rules 1–7 implement a quite simple code transformation. Nevertheless, they demonstrate an approach to code transformation with rewriting rules that is also useful in more complex scenarios. Such transformations follow the same generic structure:

(1) The rules walk through AST by adding marker terms to currently visited nodes and then removing these marker terms when the node has been processed. In the previous example, this approach was used twice: first in rules 2–3 for top-down tree traversal and then in rule 6 for bottom-up tree traversal.
(2) When the current node contains some useful information, it is recorded to be used in other rules. All needed parameters can be added to the marker term, as demonstrated in rule 5, where the term `Declare_MARK($var, $type)` contains variable name and type. Alternatively, these parameters can be recorded in FactsDB, as seen in rule 4 with `check($var, $type)` method. The FactsDB approach can be used when the recorded parameters are used far from the location in AST where they are discovered and so moving these parameters in the marker term is not practical. Also when there are multiple parameters that have to be collected from different AST nodes and later analyzed, it is easier to use FactsDB to record them. It is also possible to collect such items in the marker term using TermWare list, and then process the collected list using another set of TermWare rules.
(3) The rules can find locations in AST where changes should be made (such as adding, removing or updating nodes). Such changes can be made using previously recorded information. As an example, rule 7 adds a new variable declaration based on the marker term.
(4) Also at some AST locations the rules switch their behavior without changing AST. In such rules, previously recorded information is also used. As an example, rule 5 switches from top-down search for undeclared variables to bottom-up traversal that adds needed declarations. In this case, the rules use previously recorded information from FactsDB (`isUnchecked ($var)` and `inferType($var,$type)` methods). Rule 1 combines switching behavior (initiating top-down search) with changing nodes (subroutine to function).

As a more complex example, consider how the same approach can be used to remove side effects from pure functions:

(1) The transformation starts from function declaration and walks inside function body, similar to how rules 2–3 are implemented. However, in this case we will need additional rules to propagate the marker terms inside branching and loop statements.

(2) The rules record all external variables, such as arguments marked with `INTENT(IN)` or variables in common blocks.
(3) When the rules find a statement that modifies such external variables, the statement is removed and also recorded in FactsDB. All instances of external variable inside the pure function body are changed to an output variable with `INTENT(OUT)`.
(4) The rules search for all the calls of the pure function. After each such call, a new statement is added that reads the value from function output and performs the previously recorded operation with this value.

7 Choreography of Actors

The choreography of web services [13, 19] is a mechanism which allows describing distributed process from the perspective of individual participants rather than the deterministic sequence of imperative instructions. Due to the concurrent nature of distributed system, it is hard to model it with the sequential flows and that is why Choreography is the best description tool for the asynchronous systems. As a part of this chapter, the concept of Choreography was applied to the actor model to define processes via the description of the behavior of individual actors. In essence choreography of actors is used to define the execution plan of the program. Usage of choreography can reduce the overhead on process communication and remove a single point of failure in comparison with the synchronous sequential models, as discussed in the papers [12, 16].

Choreography defines the execution plan of the distributed process through the communication protocol for each actor. Choreography is defined as 7-tuple $<R, M, C, A, O, S, P>$, where

- R is a set of roles $[r_1 < c_1 > , r_2 < c_2 >, .., r_k < c_k >]$. Every role has predefined input channel $c_i \in C$.
- M is a set of message types $[m_1, m_2, .., m_t]$. It defines the universe of possible interactions between actors.
- C is a set of channels $[c_1 < M_{c1} > , c_2 < M_{c2} >, \ldots, c_k < M_{ck} >]$. Every channel is characterized by the subset of message types it can deliver $M_c \subseteq M$.
- O is a set of operations.
- S is a set of states of the actor.
- P is a protocol of communication, defined as $f\colon R \times M \times S \rightarrow M \times S$. For every role R it assigns a set of rules, defined as $< r_i, [< c_{in}, m_{in}, o_i, m_{out}, c_{out}] >$ where $r_i \in R$, $c_{in}, c_{out} \in C$, $m_{in}, m_{out} \in M$, $o_i \in O$.

Usage of the roles allows supporting heterogeneity for actors and computing instances. For example, if a certain role requires more intensive computations, an actor which plays that role should be allocated on a more powerful computing instance. Different strategies can be taken to optimize the allocation of actors. If all resources are homogeneous, the actor could be assigned all roles – in this case it will process incoming messages in the order of appearance in the global channel.

Channels are mapped to the roles. If an actor plays a certain role, it should listen to the input channel which corresponds to the role. If the actor plays multiple roles, it should listen to all the channels mapped to those roles. Every message in the channel should be from the subset M_c of M, which is defined as a set of messages allowed by the channel. Protocol of communication defines the expected output message of the actor in response to the incoming message. Protocol of communication allows verification of the concurrent process: if actor sends an unexpected message as an output, the system transitions to the exceptional state. The trace of all events also can be used to verify the possible global states of the distributed process. The mapping between message types and channels is a part of message type definition.

8 Scaling Individual Actors on Example of LU Factorization

During the implementation of individual actors, one of the questions to address is scalability of the computations performed by each actor. It will allow addressing the third problem of scalability – whether the data model of the actor allows distributing its computation to multiple processors. Scalability of the actors will be demonstrated on the example of matrix LU factorization, which is a common task in scientific computations in general and particularly it is used in subroutine `hcore` of QuPoly program discussed in this chapter. Therefore, initially it was modeled as an actor activated by `hcore`. However, the factorization process by itself can be modeled as a collaboration of multiple actors.

As LU factorization is a relatively small distributed program, Choreography of actors can be demonstrated on its example. We are using Block LU factorization algorithm described in the paper [20] and initially implemented as a part of LAPAC package for Fortran on MPI [21]. The algorithm is used in the cases when an operation needs to be executed on a large matrix which does not fit into the memory of single computation instance. The algorithm splits the large matrix into square blocks and distributes these blocks between all available computing instances. Algorithm expresses LU factorization of the large matrix as a sequence of transformations applied to the matrix blocks. Mathematical expressions (5), (6), (7) defines the transformations of the matrix blocks, applied recursively during execution of the algorithm.

$$A_{00} = L_{00} * U_{00} \tag{5}$$

$$A_{10} = L_{10} * U_{00}; A_{01} = L_{00} * U_{01} \tag{6}$$

$$A_{11}^{'} = A_{11} - L_{10} * U_{01} \tag{7}$$

Every iteration of the algorithm is applied to the top left unprocessed matrix block A_{00} on the matrix diagonal. Expression (5) represents the block matrix A_{00} as a factor of lower-trigonal L_{00} and upper-trigonal U_{00} matrices, which are calculated using the sequential algorithm for the LU matrix factorization. Expression (6) defines the transformation applied to the block matrix column A_{10} and block matrix row A_{01}, which allows calculating lower triangular block matrix column L_{10} and upper triangular block matrix row U_{01}. Expression (7) defines the update done to the remaining A_{11} blocks of the matrix in order to retrieve matrix A'_{11}. The process is repeated on the matrix $A'_{11,}$ taking the top left block matrix of A'_{11} as an A_{00}. Choreography for this algorithm is defined as follows:

```
Messages: <MatrixBlockReady>
Operations: <splitMatrix, processA00, multiplyA10,
  multiplyA01, readU01, updateA11>
Roles: <LUFactorizer>
States: <waiting>
Channels:
<A00, A01, A10, A11, L00I, L10, L10U01, U00I, U01,
calculateA11, A11calculated, terminate, init>
Protocol: [
  <LUFactorizer, [
    <waiting, init, splitMatrix, [A00, A01, A10, A11],
    waiting >,
    <waiting, A00, processA00, [U00I, L00I, terminate],
    waiting >,
    <waiting, U00I, multiplyA10, L10, waiting>,
    <waiting, L00I, multiplyA01, U01, waiting>,
    <waiting, L10, readU01, [L10, U01, L10U01], waiting>,
    <waiting, L10U01, updateA11, [A00, A01n A10m, A11],
    waiting>,
    <waiting, terminate, stopActor, Nil, waiting>]>]
```

In the initial state *s0*, all the channels of choreography are empty, and the set of actors is empty. Execution of the choreography starts with activation of the actors and assigning roles to them. The number of actors defines the scale factor of the system. For example, if matrix factorization should be distributed between 4 computing instances, 1 actor should be started on each instance and role `LUFactorizer` assigned to it. Activation of the actors for choreography usually is performed by the parent node in the actors' graph. For the new actors to start working, channels should contain messages to process. Parent actor node sends the first message to the `Init` channel. Choreography defines that init message should be processed by any actor with the `LUFactorizer` role. It should invoke `splitMatrix` operation and send messages to the `A00, A01, A10, A11` channels, where `splitMatrix` operation splits the large matrix into the blocks. Messages in output channels are created by the actor for each matrix block extracted from the large matrix and placed in the storage. When the message appears in the `A00` channel, the first available actor playing `LUFactorizer` role will pick that message, execute operation `processA00` which will perform LU factorization of the `A00` block and store the result. It corresponds to the step (5) of the algorithm. Then, this actor will send messages to the `U00I` and `L00I` channels with the location of the inversed `U00I` and `L00I` parts. At this point of time, the first available actor will process the message in those channels which should correspond to the (6) step of the algorithm. After that, the messages will be sent to `L10, U01, L10U01` channels. The next step happens when actors pick up messages from the `L10U01` channel, execute operation `updateA11` which corresponds to the step (7) of the algorithm. Messages are sent to `A00`, `A01n`, `A10 m`, `A11` channels, which initiates the new iteration of the recursive call.

Synchronization of the computation process happens naturally: if there is no message available in the channel, all the actors are waiting for the message to appear. It means that there is no need for a global synchronizer if barrier synchronization needs to be implemented between the steps of the algorithm. In this example, the only role assigned to the actor is `LUFactorizer`, which means that there is no specialization between actors. This allows to maximize the throughput of the messages between actors – the first available actor processes the message from the channel and this actor becomes available again as soon as the message is processed. The delays may happen if some actors pick up the messages which take a long time to process, and all channels are empty so that other actors cannot start execution. This can be addressed with the tuning of the number of the actors or differentiation of the roles.

As we discussed earlier, every actor is allocated on a separate computing instance. When a certain matrix block is needed for execution of the actor's operation, the actor performs network connection to the storage where the block is located and loads it to local storage before performing the operation. If the actor already has the corresponding piece in the local storage, it can perform the operation without spending additional time on data transfer. This is called data locality. It was stated earlier that the first available actor loads the message from the channel, which does not take into account data locality when distributing works between workers. There are multiple optimizations which can improve data locality: (a) usage of fast shared storage to minimize overhead on data load; (b) separation of the matrix into smaller blocks to minimize idling time while matrix block is loaded; (c) equal distribution of matrix blocks between computing nodes to maximize the probability that the block is located close to the node where the operation is executed.

9 Changes to Fortran Program to Support Choreography and Cloud Infrastructure

In the cloud infrastructure, actors are deployed to the computing instances, where each instance might have one or more actors deployed. Every actor is a service running in the background and executing the operations based on the messages available in the channels. In the cloud infrastructure, computing instances are usually separate machines which are running either on bare metal boxes or virtualized environment. Cloud platform operation system, such as CloudStack, provides APIs to do scaling – provisioning of the nodes with the predefined configuration on demand. This capability can be used by the actor which activates new actors. If every activated actor needs separate computing instance to execute, this operation will be invoked during activation. Otherwise, the actor will be activated on existing machines. As an example, in the case of LU factorization, we used a stable number of nodes – 4 of the same type, but the number of actors on each node varied, depending on the length of the channels. This allows utilizing computing resources in an optimal way. The number of nodes, the time they are actively working, and the size of the used memory affects the cost of computation. To minimize the cost, the optimal parameters of configuration should be chosen. A similar approach can be used for minimization of the cost as the one proposed for performance optimization of the service-oriented program based on load estimation [11].

In the cloud infrastructure, channels are implemented as message queues: FIFO channels with the guaranteed delivery. Amazon SQS service provides message queue as a service, but another option can be an installation of RabbitMQ, Kafka or similar platform on the computing instance to be used as a messaging channel (Fig. 2).

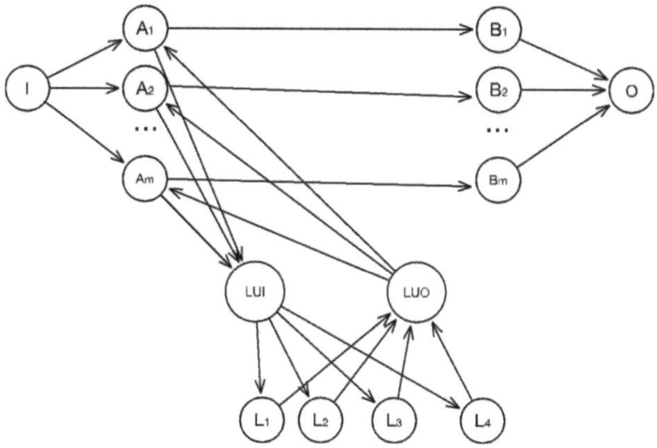

Fig. 2. Actor model applied to actor's subroutine.

After Fortran program is converted to the pure functions, every function can run as a separate program. To fulfill the second requirement of the actor model to use the message passing for communication, these programs should be converted to the actors. For this, we implement AbstractActor library using Java. Its job is to follow the choreography based on the description of the choreography protocol: read messages from the input channels, invoke operations, send messages to the output channels. AbstractActor library implements a generic mechanism of choreography and takes protocol of the choreography as an input. It invokes Fortran functions as a local program that executes operations defined in the protocol. In this way, every pure function of Fortran is wrapped in the AbstractActor which is able to follow the protocol of execution. Part of the AbstractActor implementation is the data retrieval step as a preparation for the operation execution. If data does not exist on a local machine where the actor is running, it will get the address of the data from the input message and download the data using the Linux `rsync` command.

10 Evaluation of the Approach

For the evaluation of the approach with the Choreography of actors, we use Block LU factorization example with a large matrix as an input. The testing environment consists of the 8 t2.medium Amazon AWS EC2 instances (2 CPUs, 4 Gb RAM, EBS). The SQS queue service is used to exchange notification messages between actors. The experiment consists of 3 parts:

- Measurement of scalability, according to the definition (1) is performed on the matrix of 12 × 12 blocks, 1000 × 1000 each (Fig. 3a). During the experiment, as the number of the involved instances is increased, the computation time for parallel implementation decreases almost linearly, demonstrating the possibility of the system to involve additional resources. Additionally, scale factor close to 2 is measured for all number of instances (Fig. 3b).

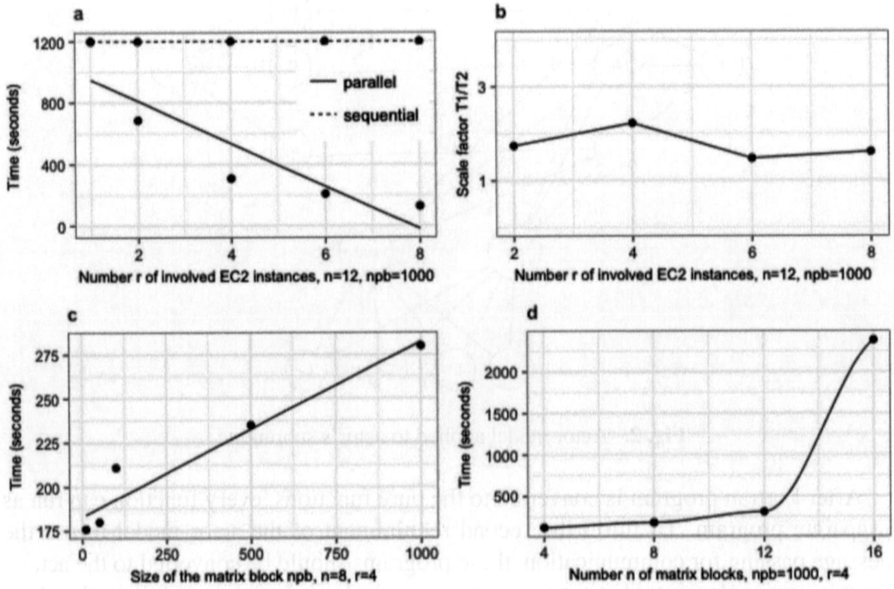

Fig. 3. **(a)** Time needed to factorize the matrix of 12 × 12 blocks, 1000 × 1000 each with variable number *r* of EC2 instances. It combines two experiments: top line is the sequential program, bottom line is the distributed program. Demonstrates the linear decrease of computing time with the growing number of instances. **(b)** Scale factor for factorization of the matrix of 12 × 12 blocks, 1000 × 1000 expressed as a ratio of time *T1* needed on the number of EC2 instances *r1* to time *T2* on twice as many instances *r2*. Demonstrates constant scale factor ~ 2 equal to *r2/r1* **(c)** Time needed to factorize matrix 8 × 8 blocks on the same number of EC2 instances $r = 4$, with the variable size of the block. Demonstrates the time increased due to computation cost with no additional losses on communication **(d)** Time needed to factorize matrix of square blocks 1000 × 1000 each, with variable number of blocks, on same number of EC2 instances. Demonstrates the cost of the actors communication.

- Measurement of the time needed for the program to factorize the matrix with the same number of matrix blocks *n*, but growing size of the block *npb* (Fig. 3c). This characterizes the time performance of algorithm without the synchronization overhead. On small matrices that fit entirely in the instance memory with *npb* from 10 to 1000, the time grows almost linearly.

– Measurement of the time needed for the program to factorize the matrix with the growing number of blocks *n*, but the same size of the blocks *npb* (Fig. 3d). It characterizes the complexity of synchronization of the actors, with the growing amount of message exchange.

The experiment data shows that the actor model can effectively adapt to the additional computation resources and allows linear scalability. Besides, with the same number of involved computation instances and growing size of the data the time for computation grows almost linearly with the small number of matrix blocks. However, with the growing number of matrix blocks processed in parallel by actors, time complexity grows as the quadratic function, which highlights inefficiencies with the actors' synchronization. This is partially explained by the algorithm of factorization, selected for the experiment, which has operations on the critical path that can be performed only synchronously and partially by the configuration of the channels for communication between actors. This can be optimized by proper selection of the roles for the actors which varies in different applications.

11 Related Work

The problem of transforming legacy applications for cloud environments is extensively studied in the literature. Paper [22] provides a systematic literature review of 23 studies related to migrating legacy applications into cloud. Three major aspects of migration are considered: 1) Migration planning; 2) Migration execution; 3) Migration evaluation. Our approach mostly centers on migration execution, although step 3 (mentioned in Sect. 3) deals with migration planning, and step 7 with migration evaluation. It is worth noting that, out of 23 reviewed studies, authors of [22] identified 9 studies that attempt automating migration process to some degree. Out of these 9 studies, 5 studies attempt fully-automated approach and 4 studies attempt semi-automated approach. However, automation attempts usually focus on planning and evaluation tasks, such as selection of cloud provider, deployment and configuration of cloud services, validation of various requirements after transformation. The transformation step itself usually requires manual input, which is comparable to our approach. The benefit of our approach is capturing transformations as formal rewriting rules. Such rules still need to be developed when a new transformation is attempted for the first time. Once developed, the rules can be reused and/or analyzed, improving the productivity of developer and performance of transformed code.

Many studies, including CloudStep [23] and CloudDSF [24], propose decision frameworks for choosing various aspects of cloud migration process. Such frameworks provide step-by-step guidelines for choosing cloud provider, scalability options, conditions that trigger changes in the number of instances, etc. In a similar way, CloudGenius [25] provides an automated tool for choosing the best combination of cloud provider and VM image formats. As mentioned in Sect. 3, this chapter does not focus on such concerns – instead, we have selected a popular Amazon EC2 provider and perform transformation for this target platform. In future, it is possible to combine our approach with such decision frameworks to provide a more comprehensive solution for legacy code transformation.

High-level models of legacy applications are used in CloudMIG approach [26]. This semi-automated approach builds models of legacy application and applies heuristic rules to choose the most suitable migration strategies. The difference from our approach is that CloudMIG model focuses on more coarse-grained application components, while our approach works on the level of individual subroutines and statements in source code. In this aspect, our approach is close to Pydron [6] that provides code-centered semi-automated parallelization by annotating source code (but without high-level models, Pydron is more focused on currently supported code rather than legacy code where original developers are frequently unavailable). Our more detailed models allow automating certain steps of re-engineering on source code level, while CloudMIG relies on manual re-engineering once the optimal strategies were selected automatically. On the other hand, so far we have only tried our approach on small and medium-sized legacy codebases. Once we move to larger and more complex legacy applications, we anticipate the need for such component-level models, perhaps combining our approach with those similar to [26].

Challenges of adapting scientific computations to distributed cloud environment are discussed in the paper [27] for Kepler execution environment. Kepler supports many execution models and platforms, including MPI, Hadoop clusters, Grid and Cloud environments. Depending on the target platform, either orchestration (director model) or choreography (actor model) is used. Bio-Cirrus [7] also provides facilities for executing bioinformatics tasks on MapReduce platforms. The advantage of these environments is that they already support many popular scientific applications. However, when attempting to run smaller and less-known applications, they should be adapted manually to execute on target platforms. We hope that our approach will allow us to capture generic transformation rules that would be applicable to many different applications. Our future research directions include validating this hypothesis.

Our choice of actor model for implementing service choreography is supported by a number of studies, such as CoDE framework [28], KOALA approach [29] and already mentioned Kepler project [27]. The difference of our approach is that the transformation of legacy code into actors is captured in form of rewriting rules, allowing the reuse of such transformations.

12 Conclusion

The chapter describes the work in progress of scaling legacy Fortran code using cloud platforms. Proposed architecture uses choreography of actors which allows to implement scalable distributed system and reduces overhead on message passing. Using rewriting rules technique allows to capture and reuse transformations of legacy code to modern distributed platforms. The experiment is performed on application from quantum chemistry field for calculation of atoms orbitals. One of the main results of the chapter is a methodology for adjustment of the legacy source code to the cloud infrastructure, including transition steps to distributed scalable architecture.

Our future research directions include automating additional transformation steps using TermWare framework, applying our approach to different applications, as well as testing different cloud configurations to find the most efficient ways of parallelizing legacy applications.

References

1. PGI Compilers & Tools. http://www.pgroup.com/products/pvf.htm
2. High Performance. http://hpff.rice.edu
3. Fortran is more popular than ever; Intel makes it fast. https://software.intel.com/en-us/blogs/2011/09/24/fortran-is-more-popular-than-ever-intel-makes-it-fast
4. Coarrays in the next Fortran Standard. ftp://ftp.nag.co.uk/sc22wg5/N1751-N1800/N1787.pdf
5. Netlib Repository. http://netlib.org
6. Müller, S.C., Alonso, G., Amara, A., Csillaghy, A.: Pydron: semi-automatic parallelization for multi-core and the cloud. In: 11th USENIX Symposium on Operating Systems Design and Implementation (OSDI 2014), pp. 645–659 (2014)
7. Karlsson, T.J.M., Tirado, Ó.T., Ramet, D., Lago, J., Cano, J.F., Chelbat, N., Trelles, O.: Bio-Cirrus: a framework for running legacy bioinformatics applications with cloud computing resources. In: Rojas, I., Joya, G., Cabestany, J. (eds.) IWANN 2013. LNCS, vol. 7903, pp. 200–207. Springer, Heidelberg (2013). doi:10.1007/978-3-642-38682-4_23
8. Doroshenko, A., Khavryuchenko, V., Iegorov, V., Suslova, L.: Modeling for quantum chemistry computations (in Russian). Upravliayushchie Sistemy i Mashiny **5**, 83–87 (2012)
9. Doroshenko, A., Shevchenko, R.: A rewriting framework for rule-based programming dynamic applications. Fundam. Inf. **72**, 95–108 (2006)
10. Tulika, E., Zhereb, K., Doroshenko, A.: Fortran programs parallelization using rewriting rules technique (in Ukrainian). Prob. Program. **2–3**, 388–397 (2012)
11. Tulika, E.: Performance optimization in SOA using load estimation and load balancing (in Ukrainian). Prob. Program. **2–3**, 193–201 (2010)
12. Barker, A., Weissman, J.B., van Hemert, J.I.: Reducing data transfer in service-oriented architectures: the circulate approach. IEEE Trans. Serv. Comput. **5**, 437–449 (2012)
13. Peltz, C.: Web services orchestration and choreography. Computer **36**, 46–52 (2003)
14. Golub, G.H., Ortega, J.M.: Scientific Computing: An Introduction with Parallel Computing. Elsevier, Amsterdam (2014)
15. Schelter, S., et al.: Distributed matrix factorization with MapReduce using a series of broadcast-joins. In: Proceedings of the 7th ACM Conference on Recommender Systems, pp. 281–284. ACM (2013)
16. Barker, A., Walton, C.D., Robertson, D.: Choreographing web services. IEEE Trans. Serv. Comput. **2**, 152–166 (2009)
17. Gaspari, M., Zavattaro, G.: An algebra of actors. In: Ciancarini, P., Fantechi, A., Gorrieri, R. (eds.) FMOODS 1999. ITIFIP, vol. 10, pp. 3–18. Springer, Boston, MA (1999). doi: 10.1007/978-0-387-35562-7_2
18. Kshemkalyani, A.D., Singhal, M.: Distributed Computing: Principles, Algorithms, and Systems. Cambridge University Press, Cambridge (2011)
19. Su, J., Bultan, T., Fu, X., Zhao, X.: Towards a theory of web service choreographies. In: Dumas, M., Heckel, R. (eds.) WS-FM 2007. LNCS, vol. 4937, pp. 1–16. Springer, Heidelberg (2008). doi:10.1007/978-3-540-79230-7_1
20. Demmely, J.W., Highamz, N.J., Schreiberx, R.S.: Block LU Factorization (1992)

21. Choi, J., et al.: Design and implementation of the ScaLAPACK LU, QR, and Cholesky factorization routines. Scientific Programming. **5**, 173–184 (1996)
22. Jamshidi, P., Ahmad, A., Pahl, C.: Cloud migration research: a systematic review. IEEE Trans. Cloud Comput. **1**, 142–157 (2013)
23. Beserra, P.V., et al.: Cloudstep: a step-by-step decision process to support legacy application migration to the cloud. In: 2012 IEEE 6th International Workshop on the Maintenance and Evolution of Service-Oriented and Cloud-Based Systems (MESOCA), pp. 7–16. IEEE (2012)
24. Andrikopoulos, V., Darsow, A., Karastoyanova, D., Leymann, F.: CloudDSF – the cloud decision support framework for application migration. In: Villari, M., Zimmermann, W., Lau, K.-K. (eds.) ESOCC 2014. LNCS, vol. 8745, pp. 1–16. Springer, Heidelberg (2014). doi: 10.1007/978-3-662-44879-3_1
25. Menzel, M., Ranjan, R.: CloudGenius: decision support for web server cloud migration. In: Proceedings of the 21st International Conference on World Wide Web, pp. 979–988. ACM (2012)
26. Frey, S., Hasselbring, W.: The CloudMIG approach: model-based migration of software systems to cloud-optimized applications. Int. J. Adv. Softw. **4**, 342–353 (2011)
27. Płóciennik, M., et al.: Approaches to distributed execution of scientific workflows in Kepler. Fundam. Inf. **128**, 281–302 (2013)
28. Bergenti, F., Poggi, A., Tomaiuolo, M.: An actor based software framework for scalable applications. In: Fortino, G., Di Fatta, G., Li, W., Ochoa, S., Cuzzocrea, A., Pathan, M. (eds.) IDCS 2014. LNCS, vol. 8729, pp. 26–35. Springer, Cham (2014). doi:10.1007/978-3-319-11692-1_3
29. Wavresky, F., Lee, S.-W.: A methodology towards the adaptization of legacy systems using agent-oriented software engineering. In: Proceedings of the 31st Annual ACM Symposium on Applied Computing, pp. 1407–1414. ACM (2016)
30. Doroshenko, A., Zhereb, K., Yatsenko, O.: Developing and optimizing parallel programs with algebra-algorithmic and term rewriting tools. In: Ermolayev, V., Mayr, H.C., Nikitchenko, M., Spivakovsky, A., Zholtkevych, G. (eds.) ICTERI 2013. CCIS, vol. 412, pp. 70–92. Springer, Cham (2013). doi:10.1007/978-3-319-03998-5_5
31. Doroshenko, A., Zhereb, K.: Parallelizing legacy fortran programs using rewriting rules technique and algebraic program models. In: Ermolayev, V., Mayr, H.C., Nikitchenko, M., Spivakovsky, A., Zholtkevych, G. (eds.) ICTERI 2012. CCIS, vol. 347, pp. 39–59. Springer, Heidelberg (2013). doi:10.1007/978-3-642-35737-4_3
32. Doroshenko, A., Khavryuchenko, V., Tulika, E., Zhereb, K.: Transformation of the legacy code on Fortran for scalability and cloud computing (in Ukrainian). In: 10th International Conference of Programming UkrPROG 2016, pp. 133–140. CEUR-WS (2016)

Descriptive Models of System Dynamics

G.N. Zholtkevych[1(✉)], K.V. Nosov[1], Yu.G. Bespalov[1], L.I. Rak[1], E.V. Vysotskaya[2], Y.B. Balkova[1], and V.K. Kolomiychenko[2]

[1] V. N. Karazin Kharkiv National University, 4, Svobody Sqr., Kharkiv 61022, Ukraine
g.zholtkevych@gmail.com, k.nosov@karazin.ua
[2] Kharkiv National University of Radio-Electronics, Science ave., 14, Kharkiv 61166, Ukraine

Abstract. Nowadays, while performing their investigations, researchers often face large datasets requiring fast processing for analysis and drawing adequate conclusions. Data mining, statistical methods and big data analytics provide an impressive arsenal of tools allowing scientists to address those tasks. However, in some cases investigators need techniques enabling on the base of relatively simple and cheap measurements of easily accessible parameters to build useful and meaningful concepts of an area of research.

In our paper two classes of dynamical models aimed at revealing between-component relationships in natural systems with feedback are presented. The idea of the both models follows from the frameworks of theoretical biology and ecology, where pairwise interactions between the parts of a system are regarded as background of the system's behavior. Both deterministic and stochastic cases are considered, that allow us to determine the direction of pairwise relationships in deterministic case and the direction and strength of relationships in stochastic one.

Keywords: Dynamical systems · Deterministic models · Stochastic models · Markov chains

1 Introduction

The world of scientific research has being immersing into an extraordinary information explosion over past decades, accompanied by the rapid growth in the use of Internet and the number of connected computers worldwide. We see a rate of increase in data growth that is faster than at any period throughout history. Enterprise application and machine-generated data continue to grow exponentially, challenging experts and researchers to develop new innovative techniques to evaluate hardware and software technologies and to develop new methods of big data study.

The problems raise during data acquisition and preliminary exploration, when the amount of data requires us to make decisions, often in an *ad hoc* manner, on importance and interpretability of data. Besides, much data today

A. Ginige et al. (Eds.): ICTERI 2016, CCIS 783, pp. 97–114, 2017.
https://doi.org/10.1007/978-3-319-69965-3_6

are not natively in structured format, have gaps and are incomplete. Hence, data analysis, organization, retrieval, and modeling are foundational challenges. Finally, presentation of the results and its interpretation by non-technical domain experts is crucial to extracting actionable knowledge.

Our study is devoted to a well-known problem of revealing conditions of stability in natural systems providing long and steady development and existence of systems. Today, there is a large amount of online big data collections comprising datasets taken from different branches of biology, health sciences, ecology etc. As examples we can mention Data Centre of International Council for the Exploration of the Sea (includes hundreds of thousands marine biology related datasets) and PhysioNet at Massachusetts Institute of Technology, a huge collection of datasets of diverse physiologic signals and open-source software for study of such data.

The problem of homeostasis and stability in the living organisms community or natural systems is closely related to the problem of dynamic stability. The practical aspect of this problem is connected to the disturbance in stability of systems, that is often accompanied, for example, by outbreaks in number or biomass of species.

The study of stability in communities or natural systems is closely connected to investigation of relationships that determine the dynamic features of a system, i.e. relationships between system's parameters having influence on the system dynamics.

For decades systemic methods, for example, based on the Shannon index of diversity, have been used for studying the relationships between the structure and stability of systems. Generalizing many such approaches, Margalef [1] states that "the ecologist sees in any measure of diversity an expression of the possibilities of constructing feedback systems, or any sort of links, in a given assemblage of species". Similar ideas were therefore presented in studying the structure of correlation pleiads, cluster analysis and other statistical techniques to establish such relationships for investigating similar problems.

Despite different approaches to revealing between-component relationships, in biology and ecology there is a general approach for presenting such relationships on the base of the following pairwise relationships: $(+,+)$, $(-,-)$, $(-,+)$, $(-,0)$, $(+,0)$, $(0,0)$. In such a way they usually denote pairwise, or paired, relationships between two components of a system. This means that two components interact each other according to the symbols presented in the corresponding relationship. E.g., $(-,+)$ means that the first component takes benefits by interacting with another, while the second suffers from the first. A quantitative measure of effects derived by the relationships are introduced in the corresponding sections. For the multi-component systems, this set of relationships exhausts all possible pairwise inter-component relationships categorized by the type of effect and have been thoroughly studied in biology and ecology [2–4]. Therefore, in the current paper the analysis of the relationships structure is based on the idea of regarding the objects (say, living organisms in a community) as the components of a system between which the mentioned pairwise relationships are possible.

This allows us to present the structure of relationships in an explicit form of relationships between the system's components.

It should be noted, that mentioned relationships cannot be always revealed with the help of statistical methods. For example, correlation analysis is initially used for estimation of a relationship between two variables, but it covers only statistical relation and cannot reveal a cause-effect relationship [5].

There are statistical methods (structural relation modeling, analysis of path and adjacent techniques), which are devoted to revealing between-component relationships (and other tasks as latent variables' analysis) and can be used for causality analysis [6–8,10]. But these methods express the relationships of a system in the terms of regression coefficients and not in the form of paired relationships. So interpretation of results of an analysis is occasionally difficult (e.g. while studying the relationships between feedback system and homeostasis in a community) and requires additional assumptions.

The models suggested in the paper tend to express the component relations in an explicit, easy-to-understand form based on pairwise relationships. Besides, intra-component relations are allowed. The models except the structure of relationships also reveal the dynamics of the system, deterministic for one case and probabilistic in another, that enables to observe the changes of the system's states over time. These advantages determine topicality and importance of the study presented herein.

2 Theory

Below, we present two dynamical models developed for revealing between-component relationships on the base of observations of real natural system.

First model has a deterministic dynamic, finite number of states and discrete time. As it is described in [11] at length, here we describe the model in short.

The second model has stochastic nature and will be describe in the paper in details.

Both models have a common background, so we begin with its description and later will go to specific properties of each models.

We assume that a natural system to be modelled comprises N components, which can be denoted by $A_1, A_2, \ldots, A_N$. Each component has a nature intrinsic to the system, for example, the number of animals or amount of biomass of different species etc. It is assumed that values of each component are integer numbers 1, 2, …, K, i.e. each component may be at K levels. The value 1 means a minimum amount of a component, the value K means maximum, i.e. a component value varies from 1 to K.

The system develops in discrete time and the moments of time are denoted $t = 0, 1, \ldots$. So, the value of the component A_i at the moment of time $t = 0, 1, \ldots$ are numbers $A_i(0)$, $A_i(1)$, ….

Next properties of a system are different for deterministic and stochastic cases, so we shall describe them separately.

2.1 Deterministic Model Revealing the Direction of Between-Component Relationships

We begin with deterministic case discussed, as mentioned, in [11] and was named the Discrete model of dynamical systems with feedback. For the deterministic system its state at the moment $t+1$ is fully determined by the state at the moment t.

If the system at the moment t is in the state $A_1(0), A_2(0), \ldots, A_N(0)$, all the following states can be written as the trajectory, where each column is a state at a corresponding moment of time:

$$\begin{pmatrix} A_1(0) & A_1(1) & A_1(2) & \ldots \\ A_2(0) & A_2(1) & A_2(2) & \ldots \\ \vdots & \vdots & \vdots & \\ A_N(0) & A_N(1) & A_N(2) & \ldots \end{pmatrix}. \tag{1}$$

In the theory of dynamical systems [12], such a system is called a free dynamical system with discrete time. The system has only finite number of states, so there exists a positive integer $\mathcal{T}$, called a period of the trajectory, for which the conditions of periodicity hold

$$\begin{pmatrix} A_1(s) \\ A_2(s) \\ \vdots \\ A_N(s) \end{pmatrix} = \begin{pmatrix} A_1(s+\mathcal{T}) \\ A_2(s+\mathcal{T}) \\ \vdots \\ A_N(s+\mathcal{T}) \end{pmatrix},$$

for enough large s.

Taking into account the periodicity, we extract the minor

$$\begin{pmatrix} A_1(s) & A_1(s+1) & \ldots & A_1(s+\mathcal{T}-1) \\ A_2(s) & A_2(s+1) & \ldots & A_2(s+\mathcal{T}-1) \\ \vdots & \vdots & \ddots & \vdots \\ A_N(s) & A_N(s+1) & \ldots & A_N(s+\mathcal{T}-1) \end{pmatrix} \tag{2}$$

from (1) presenting full description of the system's dynamics.

Now we introduce the concept of relationships between components. Let $\Omega = \{-, 0, +\}$—three-entries set. A relationship between specified components A_i and A_j is determined as an entry from the set $\Omega \times \Omega$ and denoted by $\Lambda(A_i, A_j) = (\omega_1, \omega_2)$, where $\omega_1 \in \Omega$, $\omega_2 \in \Omega$. If $\Lambda(A_i, A_j) = (\omega_1, \omega_2)$, it means that:

- if $\omega_1 = \{-\}$, then increasing the value of A_j will determine the decrease of the value of A_i.
- if $\omega_1 = \{0\}$, then the A_j doesn't influence the value of the component A_i.
- if $\omega_1 = \{+\}$, then large values of the A_j will raise the value of the A_i.

The relationship Λ is antisymmetric in the following sense: $\Lambda(A_i, A_j) = (\omega_1, \omega_2)$ implies $\Lambda(A_j, A_i) = (\omega_2, \omega_1)$. It is also assumed, that inner relationships (self-relationship $\Lambda(A_i, A_i)$) are symmetric—$(0, 0)$, $(-, -)$, and $(+, +)$ for any A_i.

Assume that all the relationships $\Lambda(A_j, A_i)$ between all pairs (A_j, A_i) of components $A_1, A_2, \ldots, A_N$ are given. For each A_j and each $(s, u) \in \Omega \times \Omega$, let $L_j(s, u) = \{A_i | \Lambda(A_j, A_i) = (s, u)\}$ (the set of components, with which A_j has the relationship (s, u)). We can express the relationships between the components by the following relationships' matrix

$$\begin{bmatrix} & A_1 & A_2 & \ldots & A_N \\ A_1 & (\omega_1, \omega_1) & & & \\ A_2 & (\omega_2, \omega_1) & (\omega_2, \omega_2) & & \\ \vdots & \vdots & \vdots & \ddots & \\ A_N & (\omega_N, \omega_1) & (\omega_N, \omega_2) & \ldots & (\omega_N, \omega_N) \end{bmatrix}. \tag{3}$$

Taking into account the antisymmetric property, the entries above main diagonal in (3) are omitted, since they can be recovered by the lower triangular part of matrix.

Let $\varkappa = \{1, 2, \ldots, K\}$ be the set of the states of the components and $N_j(s, u)$ is the number of components in the set $L_j(s, u)$, $j = 1, 2, \ldots, N$, $(s, u) \in \Omega \times \Omega$. A transition from the state $(A_1(t), A_2(t), \ldots, A_n(t))$ to the state $(A_1(t + 1), A_2(t + 1), \ldots, A_n(t + 1))$ is described by N transition functions F_j, each of which defines the mapping

$$\varkappa^{N_j(+,+)+N_j(+,0)+N_j(+,-)+N_j(-,+)+N_j(-,0)+N_j(-,-)} \mapsto \varkappa.$$

This mapping symbolically is expressed by the formula

$$\begin{aligned} &A_j(t + 1) = F_j(A_k(t) \in L_j(+, +), A_k(t) \in L_j(+, 0), \\ &A_j(t) \in L_k(+, -), A_k(t) \in L_j(-, +), \\ &A_k(t) \in L_j(-, 0), A_k(t) \in L_j(-, -)), \ j = 1, 2, \ldots, N, \end{aligned} \tag{4}$$

where $A_k(t) \in L_j(+, +)$, $A_k(t) \in L_j(+, 0)$, ... are the values of $A_k(t)$ belonging to $L_j(+, +)$, $L_j(+, 0)$, ... respectively.

The transition function, introduced by equation (4), is quite natural in its structure. Given component A_j is influenced only by those components, which indeed influence A_j, i.e. the components from the sets $L_j(+, \omega)$ and $L_j(-, \omega)$ for any $\omega \in \Omega$.

Two Types of Relationships, Intrinsic to Natural Systems. The formula (4) presents a general form of transition of the system from the state at the moment t to the state at $t + 1$.

For more detailed description of the dynamics of a natural system, numerical experiments and procedures of system identification one needs to specify explicit form of the mappings.

We introduced two approaches based on the concepts of biological interactions: the weight functions' approach and the approach based on principles of Justus von Liebich's law.

Define the following functions on the set $\varkappa$: $\mathrm{Inc}(A) = \min\{K, A+1\}$, $\mathrm{Dec}(A) = \max\{1, A-1\}$.

The system dynamics according to the weight functions' approach. First we define the type of dynamics, which takes into account the weighted sum of all $A_j(t)$ (including $A_i(t)$) for calculating the value of the component A_i at the moment $t+1$.

As we defined above, for each j ($j = 1, 2, \ldots, N$) and each pair $(s, u) \in \Omega \times \Omega$ there exists the set $L_j(s, u)$ with $N_j(s, u)$ entries. Assume that $\varphi_{j,1}^{\langle s,u \rangle}(\cdot)$, $\varphi_{j,2}^{\langle s,u \rangle}(\cdot)$, $\ldots$, $\varphi_{j,N_j(s,u)}^{\langle s,u \rangle}(\cdot)$ are to be the functions of interactions of those components, with which the A_j has relationships (s, u). The functions are defined on the discrete set $\varkappa$ and have the following properties:

1. $\varphi_{j,k}^{\langle +,+ \rangle}(\cdot)$, $\varphi_{j,k}^{\langle +,0 \rangle}(\cdot)$, $\varphi_{j,k}^{\langle +,- \rangle}(\cdot)$ are increasing functions.
2. $\varphi_{j,k}^{\langle -,+ \rangle}(\cdot)$, $\varphi_{j,k}^{\langle -,0 \rangle}(\cdot)$, $\varphi_{j,k}^{\langle -,- \rangle}(\cdot)$ are decreasing functions.
3. $\varphi_{j,k}^{\langle s,u \rangle}(1) = 0$ for any $(s, u) \in \Omega \times \Omega$.

We also introduce the numbers $\delta_j > 0$ ($j = 1, 2, \ldots, N$) which can be called thresholds of sensitivity.

For the system's state at the moment t, the following value is calculated

$$\begin{aligned} d_j = &\textstyle\sum_{A_k \in L_j(+,+)} \varphi_{j,k}^{\langle +,+ \rangle}(A_k(t)) + \sum_{A_k \in L_j(+,0)} \varphi_{j,k}^{\langle +,0 \rangle}(A_k(t)) \\ &+\textstyle\sum_{A_k \in L_j(+,-)} \varphi_{j,k}^{\langle +,- \rangle}(A_k(t)) + \sum_{A_k \in L_j(-,+)} \varphi_{j,k}^{\langle -,+ \rangle}(A_k(t)) \\ &+\textstyle\sum_{A_k \in L_j(-,0)} \varphi_{j,k}^{\langle -,0 \rangle}(A_k(t)) + \sum_{A_k \in L_j(-,-)} \varphi_{j,k}^{\langle -,- \rangle}(A_k(t)). \end{aligned} \tag{5}$$

The value of the component $A_j(t+1)$ is calculated as follows

1. if $d_j \geq \delta_j$, then $A_j(t+1) = \mathrm{Inc}(A_j(t))$;
2. if $d_j \leq -\delta_j$, then $A_j(t+1) = \mathrm{Dec}(A_j(t))$;
3. if $-\delta_j < d_j < \delta_j$, then $A_j(t+1) = A_j(t)$.

Now, the meaning of introduced transition functions can be explained in clear way. For example, the functions $\varphi_{j,k}^{\langle -,+ \rangle}(\cdot)$ ($k = 1, 2, \ldots, N_j(-,+)$) reflects the influence upon the component A_j by components in the set $L_j(-,+)$, which are related with A_j by relationship $(-,+)$. The greater the influence (i.e. the greater values of $A_i(t)$ from the set $L_j(-,+)$), the lower the values of d_j.

The influence of other components, with which A_j has other relationships, is "weighted" in similar way. If the cumulative influence of components, interacted with A_j and expressed by Eq. (5), exceeds the threshold δ_j, then the value of A_j is changed by unit.

The threshold δ_j clearly influenced the dynamics of the system in the following way: the greater δ_j, the greater absolute value of the weighted sum d_j required for overcoming this δ_j for changing the value of A_j. So if δ_j is very large, the system becomes very inert.

The dynamics based on the Liebig's law of the minimum. Next approach is based on principles of Justus von Liebich's law (Liebig's law of the minimum) and essentially differs from first approach, which is basically additive.

Assume that the system of relationships between A_1, A_2, ..., A_N is given. For defining the system's dynamics, let's introduce two constant matrices, C and C^* of size $N \times N$. The transition function is based on the following algorithm.

Suppose the system in state $(A_1(t), A_2(t), \ldots, A_N(t))$ at time t and A_j is an arbitrary fixed component. Let i runs from 1 to N, and u means arbitrary entry of the set Ω.

1. If for the current i the equality $\Lambda(A_j, A_i) = (-, u)$ holds true, assume that

$$f_i = \begin{cases} -1, & \text{if } A_i(t) \geq c^*_{ji}, \\ 0, & \text{if } c_{ji} + 1 \leq A_j(t) \leq c^*_{ji} - 1, \\ 1, & \text{if } A_j(t) \leq c_{ji}. \end{cases}$$

The specific value of u doesn't matter because only the influence on A_i from the side of A_j matters.

2. If for the current i the equality $\Lambda(A_j, A_i) = (+, u)$ holds true, assume

$$f_i = \begin{cases} -1, & \text{if } A_j(t) \leq c_{ji}, \\ 0, & \text{if } c_{ji} + 1 \leq A_j(t) \leq c^*_{ji} - 1, \\ 1, & \text{if } A_j(t) \geq c^*_{ji}. \end{cases}$$

3. If for the current i the equality $\Lambda(A_j, A_i) = (0, u)$ holds true then it is assumed that $f_i = 0$.

After the cycle termination, the sequence f_1, f_2, ..., f_N is obtained. The value $A_j(t+1)$ is calculated according to the following rule:

$$A_j(t+1) = \begin{cases} \mathrm{Dec}(A_j(t)), & \text{if } \min\limits_{1 \leq i \leq N} f_i = -1, \\ A_j(t), & \text{if } \min\limits_{1 \leq i \leq N} f_i = 0, \\ \mathrm{Inc}(A_j(t)), & \text{if } \min\limits_{1 \leq i \leq N} f_i = 1. \end{cases} \tag{6}$$

Applying this algorithm for each $j = 1, 2, \ldots, N$, the system's state at the moment $t + 1$ is calculated.

The meaning of transition from t to $t+1$ can be explained clearly. Suppose, a given component A_j has relationship $(+, -)$ with this current component A_i (see the algorithm). According to the relationship $(+, -)$, large values of A_i should decrease A_j. Indeed, according to item 1 of the algorithm, if $A_i(t) \geq c^*_{ji}$ (in other words, when $A_i(t)$ is "large enough"), $f_i = -1$ and, according to (6), A_j would decrease if $A_j(t) > 1$. Other cases of transition work in similar way.

The System Identification Based on the Data of Observation. When we deal with real data, as a rule, we don't observe their dynamics explicitly. Often real data are unordered in time in contrast to data used for time series modeling. So we don't observe any dynamism described by the relationship (3), by the trajectory (1) or the minor (2).

Usually, the result of observation is represented by the table:

$$\tilde{M} = \begin{pmatrix} C_{11} & C_{12} & \dots & C_{1B} \\ C_{21} & C_{22} & \dots & C_{2B} \\ \vdots & \vdots & \ddots & \vdots \\ C_{N1} & C_{N2} & \dots & C_{NB} \end{pmatrix}, \tag{7}$$

where columns correspond to cases and rows correspond to components (N components, B cases). We emphasize unordered character of the data above, i.e. these is no time order between the cases in the table $\tilde{M}$.

Here we describe a principle allowing to reveal the system relationships of above mentioned type on the basis of the observation table $\tilde{M}$.

This algorithm determines inter- and intra-component relationships, which are as close as possible to relationships, which form matrix (2) in some sense.

Assume that the relationships structure is given. In that case for the initial state $(A_1(0), A_2(0), \dots, A_N(0))$ and for the given sets $L_1(u, s)$, $L_2(u, s)$, …, $L_N(u, s)$, $u \in \Omega$, $s \in \Omega$ the minor (2) can be calculated. Let

$$P = \begin{pmatrix} 1 & r_{12} & \dots & r_{1N} \\ r_{12} & 1 & \dots & r_{2N} \\ \vdots & \vdots & \ddots & \vdots \\ r_{1N} & r_{2N} & \dots & 1 \end{pmatrix}$$

is to be the correlation matrix (Pearson or Spearman) between the rows of the minor (2). Also, for the table $\tilde{M}$, the correlation matrix of its rows can be calculated:

$$\tilde{P} = \begin{pmatrix} 1 & \rho_{12} & \dots & \rho_{1N} \\ \rho_{12} & 1 & \dots & \rho_{2N} \\ \vdots & \vdots & \ddots & \vdots \\ \rho_{1N} & \rho_{2N} & \dots & 1 \end{pmatrix}.$$

Introduce the measure of distance between the correlation matrices P and $\tilde{P}$

$$D(P, \tilde{P}) = \sum_{i=1}^{N-1} \sum_{j=i+1}^{N} (r_{ij} - \rho_{ij})^2. \tag{8}$$

Set the problem of minimization $D(P, \tilde{P})$ by all possible vectors of initial states $(A_1(0), A_2(0), \dots, A_N(0))$ and all allowable sets $L_j(s, u)$, $s \in \Omega$, $u \in \Omega$ for all j

$$D(P, \tilde{P}) \mapsto \min \quad \begin{matrix} \text{(by all initial states \&} \\ \text{by all allowable sets } L_j(s, u)). \end{matrix} \tag{9}$$

The meaning of this problem can be explained in the following way. Suppose, a process in some natural system is cyclical with the trajectory (2). There is no possibility to observe the dynamics of this trajectory, i.e. a full length cycle. The observations are taken from the system at random moments of time t from s

to $s + \mathcal{T} - 1$ with equal probability. When an observation is taken, the column $(A_1(t), A_2(t), \ldots, A_N(t))^T$ from (2) is attached to the table of observations. In other words, the columns of table of observations M are obtained from (2) by equiprobable choice of columns.

The stated problem means the search for such relationships between components, that the minor (2) is to be as close as possible to the table of observations regarding the measure (8).

The following theorem proved in [11] shows that this problem is well-grounded in probabilistic sense.

Theorem 1. *If the table of observations $\tilde{M}$ is obtained from the minor (2) by equiprobable choice of columns, then the Pearson correlation matrix of the observations table $\tilde{P}$ converges to the correlation matrix of minor P (in probability)*

$$\lim_{B\to\infty} \rho_{ij} = r_{ij}, \quad i = 1, 2, \ldots, N, \ j = 1, 2, \ldots, N.$$

The same result takes place for the Spearman correlation matrix as well.

2.2 Additive Stochastic Model of Between-Component Relationships

Our another model is also described by the set of components $A_1, A_2, \ldots, A_N$ taking discrete values 1, 2, …, K.

But, in contrast to the first one, the second model introduces into consideration not only direction of relationships (in fact, for the first model we considered three direction—negative, neutral, and positive), but also a strength of relationships.

The structure of relationships between the components $A_1, A_2, \ldots, A_N$ is described by the following relationships matrix

$$\mathcal{M} = \begin{pmatrix} m_{1,1} & m_{1,2} & \ldots & m_{1,N} \\ m_{2,1} & m_{2,2} & \ldots & m_{2,N} \\ \vdots & \vdots & \ddots & \vdots \\ m_{N,1} & m_{N,2} & \ldots & m_{N,N} \end{pmatrix}.$$

Any entry $m_{i,j}$ reflects the strength and direction of influence of the component A_j upon the component A_i. The direction of influence is expressed by the sign of the value $m_{i,j}$ (may be $-, 0, +$) and the strength—by modulus of $m_{i,j}$ and varies from 0 to 1. So, $-1 \le m_{i,j} \le 1$ for each i, j. The influence of the component A_i on A_j is expressed by $m_{j,i}$. It is easy to see, that the relationship between the components A_i on A_j is described by the pair $(m_{i,j}, m_{j,i})$, which is in some way similar to the relationship (ω_1, ω_2) introduced for the first model.

Now describe the dynamics of transition from the state of the system at the moment t to the state at the next moment $t + 1$, i.e.

$$(A_1(t), A_2(t), \ldots, A_N(t)) \mapsto (A_1(t+1), A_2(t+1), \ldots, A_N(t+1)).$$

As for the weight functions' approach, we assume, that the set of functions $\psi_{i,j}(\cdot)$, $(i,j = 1,2,\ldots,N)$ reflecting relationships between all pairs of components, including inner relationships, are given.

All the functions $\psi_{i,j}(\cdot)$ have the following properties:

1. $\psi_{i,j}(\cdot)$ are defined on the set $\varkappa$;
2. $\psi_{i,j}(1) > 0$;
3. $\psi_{i,j}(\cdot)$ are increasing functions on $\varkappa$.

It should be noted that the property 2 of functions $\psi_{i,j}(\cdot)$ is different from corresponding property for functions $\psi_{i,j}^{\langle s,u\rangle}(\cdot)$, which requires $\psi_{i,j}^{\langle s,u\rangle}(1) = 1$.

Also assume that a positive number δ playing the a role of threshold, is given.

Let the system be in the state $(A_1(t), A_2(t), \ldots, A_N(t))$. For each pair of indices i,j $(i = 1,2,\ldots,N,\ j = 1,2,\ldots,N)$ define the random variable $\xi_{i,j}$ as follows

$$\xi_{i,j} = \begin{cases} \psi_{i,j}(A_j(t))\mathrm{sign}(m_{i,j}) & \text{with probability } |m_{i,j}| \\ 0 & \text{with probability } 1 - |m_{i,j}|. \end{cases}$$

Then we introduce the set of N random variables

$$d_i = \sum_{j=1}^{N} \xi_{i,j},\ i = 1,2,\ldots,N.$$

Using the set $(d_1, d_2, \ldots, d_N)$, it's possible to calculate the set of probabilities (p_i^-, p_i^0, p_i^+) for each i as follows

$$\begin{aligned} p_i^+ &= P(d_i \geq \delta), \\ p_i^0 &= P(-\delta \leq d_i \leq \delta), \\ p_i^- &= P(d_i \leq -\delta), \end{aligned}$$

for each i from 1 to N. This definition implies $p_i^- + p_i^0 + p_i^+ = 1$.

For each i, the transition from the state at the moment t to the state at $t+1$ is defined by

$$A_i(t+1) = \begin{cases} \mathrm{Dec}(A_i(t)) & \text{with probability } p_i^-, \\ A_i(t) & \text{with probability } p_i^0, \\ \mathrm{Inc}(A_i(t)) & \text{with probability } p_i^+. \end{cases}$$

That is, at the moment $t+1$ the value of A_i can be increased by 1, remains the same or decreased by 1 with probabilities p_i^-, p_i^0, p_i^+ correspondingly.

Applying this rule for each i, the probabilities of transition from any appropriate state $(A_1(t), A_2(t), \ldots, A_N(t))$ can be calculated.

It can be proved, that if each row of the matrix $\mathcal{M}$ include both negative and positive entries, we obtain the Markov chain with K^N states $A_1(t), A_2(t), \ldots, A_N(t)$ $(A_i \in \varkappa)$. Besides, this chain is regular, so there a unique steady-state stochastic vector $\mathbf{w}$.

Now the reasons for the problem can be explained. We assume, that a natural system is described by this model, and the probability of being of the system in states converges to the entries of the vector $\mathbf{w}$ (as the system acts as a regular Markov chain). Using the states $A_1 A_2, \ldots, A_N$ and the steady-state vector $\mathbf{w}$, we can calculate a weighted Pearson correlation matrix [13] between the components.

Describe this step at length. All states of the system can be written in the table

$$\begin{bmatrix} A_1 & A_2 & \ldots & A_{N-2} & A_{N-1} & A_N \\ 1 & 1 & \ldots & 1 & 1 & 1 \\ 1 & 1 & \ldots & 1 & 1 & 2 \\ 1 & 1 & \ldots & 1 & 2 & 1 \\ 1 & 1 & \ldots & 1 & 2 & 2 \\ 1 & 1 & \ldots & 2 & 1 & 1 \\ \vdots & \vdots & \ddots & \vdots & \vdots & \vdots \\ K & K & \ldots & K & K & K \end{bmatrix} \tag{10}$$

having K^N rows and N columns.

Write out the steady-state vector of the Markov chain in the form

$$\mathbf{w} = (w_1, w_2, \ldots, w_{K^N}),$$

where the entry w_k corresponds to k-th state in the table (10).

Taking $\mathbf{w}$ as weights, calculate the weighted Pearson correlation matrix between the columns of the table (10). Denote such the matrix by $R_{\mathbf{w}}$.

We suppose, that the explicit dynamics of our natural system is not available. In other words, we cannot observe time series of states, but can record a state of the system at random moments of time. These observations are collected in the observation table $\tilde{M}$ having N variables and B cases (after B observations). Let the Pearson correlation matrix between rows of (7) is denoted by $\tilde{R}$.

Theorem 2. *If the observation table $\tilde{M}$ is obtained according to the way described above, then*

$$\tilde{R} \to R_{\mathbf{w}} (\textit{in probability if } B \to \infty),$$

where $\tilde{R} \to R_{\mathbf{w}}$ means entry-wise convergence.

Proof. Omitted for brevity sake.

Introduce the measure of proximity for the matrices R and $\tilde{R}$

$$D(R_{\mathbf{w}}, \tilde{R}) = \sum_{i=1}^{N-1} \sum_{j=i+1}^{N} (\tilde{R}_{i,j} - [R_{\mathbf{w}}]_{i,j})^2. \tag{11}$$

The Theorem 2 means that the sample observation matrix consistently represents a true dynamics, which is not observed explicitly. This result is used as a base for identifications of entries of the relationships matrix $\mathcal{M}$. In any investigations there is only a finite number of observations (B is finite). Therefore we can try to calculate transition probabilities of the Markov chain, that provide the best approximation of a true correlation matrix by a sample matrix in the sense of the measure (11).

According to this approach, $\mathcal{M}$ is obtained by resolving the following optimization problem

$$D(R_{\mathbf{w}}, \tilde{R}) \mapsto \min \text{ by entries } m_{i,j}.$$

In fact, we find the relationships matrix $\mathcal{M}$, which makes the modelled correlation matrix as close as possible to the observe correlation matrix.

3 Case Studies

In this section we present three examples of application of developed models to various natural and technical systems.

3.1 Factors Determining Users Activity in Social Networks

First example concerns analysis of system factors affecting activity of users of social networks playing an important role in modern culture [14,15]. The structure of relationships between the components of the system for two states of the Internet-forum on fantasy literature were calculated and compared. This comparison aimed at reveal system aspects of forum visiting in two periods. One state can be regarded as "low-performance", other as "high-performance" according to number of written fanfictions (also abbreviated as fan fics, fanfics) of visitors at the site dedicated to the cycle of novels of Joanne Rowling about Harry Potter (snapetales.com). The period of first half of December 2010 is regarded as "high-performance", the second period of the first half of December 2014 is called "low-performance". For these two periods a statistically significant difference according to Student t-test ($p < 0.05$) in overall average number of visits per day was also detected.

The fanfictions were divided into 4 categories according to their length—fanfictions of small, large, and medium size; the last, fourth category includes fanfictions not related to the novels about Harry Potter.

The following values were taken as the components of the system reflecting the authors activity

- the number of *small size* fanfictions per day related to the cycle of novels about Harry Potter (denoted by **MIN**);
- the number of *large size* fanfictions per day (**MAX**);
- the number of *medium size* fanfictions (**MID**);
- the number of fanfictions not related to the cycle of novels about Harry Potter, based on another literary works (**OTHER**).

For the "high-performance" and "low-performance" periods, the structure of relationships were built. We identified the models using the Pearson correlation matrix and the approach on the base of von Liebig law, with $K = 3$ levels of components values. The structure of relationships for both period is presented in Figs. 1 and 2 correspondingly. The graphs in the figures present the structure of relationships. Nodes of the graphs correspond to the components in the systems as defined above. Edges of the graphs with embedded ovals on them present pairwise relationships revealed in the model. For example, $\Lambda(\mathbf{MIN}, \mathbf{MID}) = (+, -)$, that is presented on the graph by the corresponding oval.

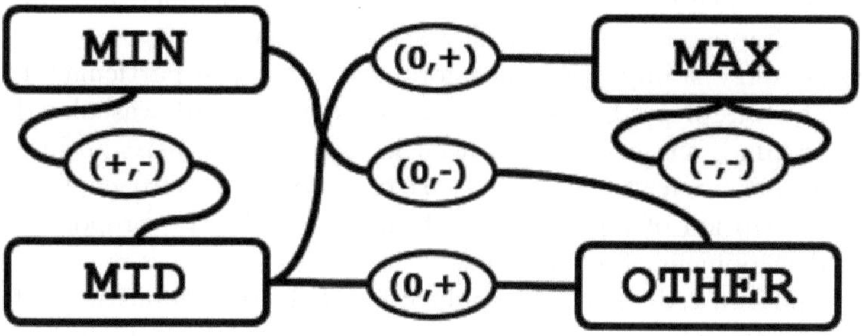

Fig. 1. The structure of relationships for "high-performance" period. Rounded rectangulars present the components of the system, the ovals present the relationships between the components.

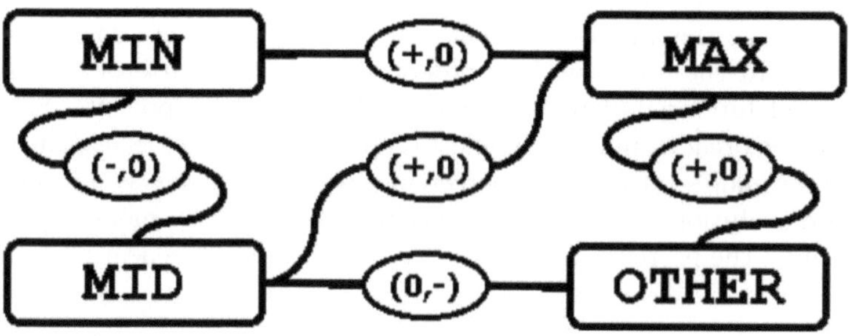

Fig. 2. The structure of relationships for "low-performance" period.

Comparing the graphs in Figs. 1 and 2 shows a system-forming role of the component **MID** for the "high-performance" period, in which **MID** positively affected the other three components. This affect disappeared in the "low-performance" period together with loss a stabilizing mechanism through the relationship $(+, -)$ between **MID** and **MIN** supporting a dynamic equilibrium of the system.

These results are consistent with empirically established ideas about significant positive role of fanfictions of medium size (**MID**) in a functioning of social networks of this category and their close relation to short-sized fanfictions

(**MIN**) representing a reaction of the most dynamic part of users. Differences in role of **OTHER** correspond to significance of "offtopic" as an index of deterioration in work of dedicated web-sites.

3.2 Relations Between the Anthropometric Parameters of Adolescents with Cardiovascular Disorders

Our next case concerns the system relations of anthropometric parameters of adolescents suffering diseases of cardiovascular system.

Anthropometry is important in school health care, in particular, for determining the factors of predisposition of adolescents to cardiovascular disorders. At the same time, among other drawbacks of currently used anthropometric methods they often refer to insufficient use of systematic approach, among other things, in description of regularities in formation of body's proportions in the individual development of adolescents.

Here we present a demo of the application of the deterministic model developed above for this purpose, built on the material of adolescents anthropometry with arterial hypertension and other forms of cardiovascular disorders. Body compositions related to overweight plays an important role in development of arterial hypertension. Taking that in account, the models for four components were built: hip circumference, waist circumference, chest circumference, and shoulder breadth divided by height of a subject. The Spearman correlation and Liebig's approach with $K = 3$ levels of components were used in modeling.

Comparison of these graphs revealed a different role of such anthropometric parameters as the hip circumference for the two group of adolescents under investigation. In the group with disorders different from arterial hypertension high values of hip circumference increase other three components. Simultaneously, shoulder breadth negatively affects hip circumference, that should form a proportion of male's future body perceiving by subconscious as harmonious on the base on evolutionary history and recognized as such by modern physiology and medicine—the proportions of male "triangle" directed beneath by edge. The

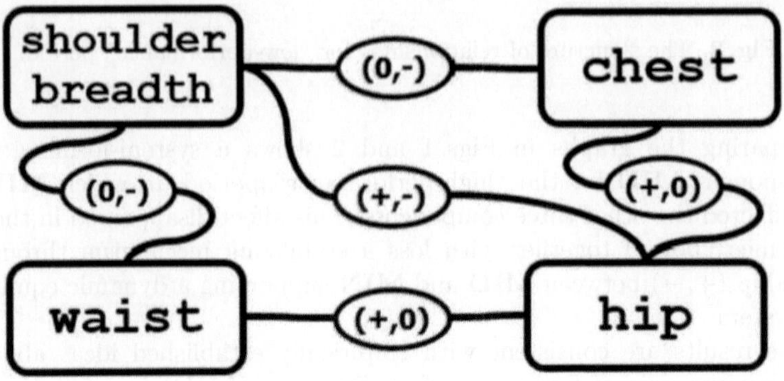

Fig. 3. The structure of relationships for adolescents arterial hypertension.

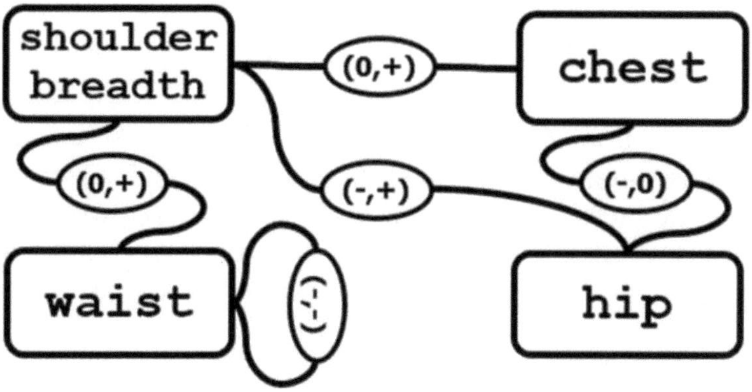

Fig. 4. The structure of relationships for adolescents without arterial hypertension.

structure of relationships in the group with hypertension prevents the formation of such a standard and associates with the accumulation of a depot fat in certain parts of a human body: relatively high values of the hip circumference negatively affects shoulders breadth and chest circumference, not directly affecting waist circumference, on which shoulders breadth positively influences.

These results, regarded by authors as preliminary, do not contradict known facts about the impact of anthropometric parameters on the risk of development of hypertension in adolescents groups.

3.3 Factors Influencing the Efficiency of Industrial Fishery at North Sea

Our last case concerns the issues on industrial fishery of Atlantic cod (*Gadus morhua*) at North Sea. The fishery of the cod plays an important role in the economy of several countries and provokes considerable interest to using mathematical models in industrial ichthyology describing large fluctuations of catching [18] (well-known example of this kind is collapse of the Atlantic northwest cod fishery in 1992).

As the demo the additive stochastic model of relationships structure between dimensional parameters of cod populations was considered. The average fish body length (**L**), the difference between Upper Length Bound and Lower Length Bound (**vL**), the average stomachs weight (**M**), and the average weight of preys of cod (**dM**) were taken as components of the model. Additive stochastic models were built according to data of International Council for the Exploration of the Sea for two years (1984 and 1989) preceding to rapid changes of CPUE (the catch per unit effort). We used the model with $K = 4$ levels of components values. The results are shown in Table 1.

In the matrix corresponding to 1984, which precedes significant (till 1990) decrease of catching yield, there are large (above 0.85) negative effects of high values of **vL** on **M** and **dM**. That is, increasing the diversity of dimensional

Table 1. Structure of relationships between dimensional parameters **L**, **vL**, **M**, and **dM** of the model for years 1984 and 1989.

1984					1989				
	L	vL	M	dM		L	vL	M	dM
L	−0.936	−0.379	0.893	0.581	**L**	−0.843	0.137	0.953	0.059
vL	0.325	−0.737	−0.405	0.519	**vL**	0.963	−0.721	−0.997	0.494
M	−0.184	−0.868	0.969	−0.081	**M**	−0.882	−0.054	0.788	0.224
dM	0.016	−0.941	0.999	−0.028	**dM**	0.091	0.066	0.941	−0.995

characteristics of the cod population, that improves the consumption possibilities of forage reserve by the cod, leads to exhaustion of food resources (reducing the number of available preys) and deterioration of preys quality (reducing the average size of forage organisms), and results in deterioration of food supply of the cod, that lowers the values of **M** and **dM**.

In the matrix corresponding to 1989, which precedes sharp increase of CPUE, recorded a year later, in 1991, there exist extremely small (below 0.07) negative effects of high values of **vL** on **M** and **dM**. In this case, the increasing diversity of sizes, that enhances abilities of consumption of forage reserve, does not lead to exhaustion and deterioration of the latter. This result of modeling explains differences described above in the dynamics of catching in accordance with modern concepts of industrial ichthyology.

Presented results bring hope for the possibility of developing methods of forecasting the cod catching with use of the stochastic models, built on the base of actual material on the size structure of the population. The data of material can be obtained among others by remote methods with the help of low-cost means and relatively little effort, and even from commercial reports.

4 Conclusion

In the paper we followed the established framework in model development, appropriated for natural sciences. Typical approach in development, among others, comprises the data selection, specification of assumptions and simplifications, selection of a mathematical modeling framework, estimation of parameter values, model diagnostics, model validation, model refinements and model application. It's clear, that all these stages of building mathematical models for biological systems are too complicated, but the most difficult task among them is the model parameters' estimation for identifying structure in the underlying biological networks.

The models presented in the paper are created for description of biological and ecological systems, based on pairwise relationships characterized by the direction (positive, negative, or neutral) for the both models and by strength varied from 0 to 1 in the stochastic model only.

The problem of parameters estimation is a true challenging problem for both models and requires development of special algorithms of numerical optimization. For example, if the system has N components and the number of levels is to be assumed K, for the first deterministic model the number of initial states is equal to K^N and the number of possible relationships' structures is equal to 3^{N^2}. For solving the stated optimization problem (9), one should built the minor (2) with use of an initial state and a relationships' structure, calculate correlation matrix P and calculate the distance (8). So, the exhaustive search of both initial states and relationships' structures in total gives us $K^N 3^{N^2}$ variants, that is a huge number for even moderate N and K.

The case studies presented in the paper, considered by the authors as preliminary and illustrating, offer the prospects of applications of proposed models.

The results of modeling of system aspects in anthropometry of adolescents present the approaches to use of this simple and cheap method for identifying the risk groups of the progress of arterial hypertension. These approaches may be applied in school medicine and, if necessary, in extreme situations for mass screening as well.

The study of system factors of performance of the web-site dedicated to fiction about characters from original works about Harry Potter, due to use of components of the system, that are invariant to the content of the web-site, may have a broader meaning in analysis of the social networks performance.

The model of the cod population as a whole does not contradict known facts on the role of fish size and state of a forage reserve in the population dynamics. At the same time, these results reveal some promises and can be used in the development of approximate methods for prediction of populations of commercial fish with use of relatively simple and inexpensive methods of data acquisition, even with use of commercial reports concerning the assortment of fish products.

References

1. Margalef, R.: Perspectives in Ecological Theory. Chicago University Press, Chicago (1968)
2. Lidicker Jr., W.Z.: A clarification of interactions in ecological systems. Bioscience **29**(8), 475–477 (1979)
3. Abrams, P.A.: On classifying interactions between populations. Oecologia **73**(2), 272–281 (1987)
4. Wootton, J.T., Emmerson, M.: Measurement of interaction strength in nature. Ann. Rev. Ecol. Evol. Syst. **36**, 419–444 (2005)
5. Shipley, B.: Cause and Correlation in Biology. Cambridge University Press, Cambridge (2004)
6. Spirtes, P., Glymour, C.N., Scheines, R.: Causation, Prediction, and Search. MIT Press, Cambridge (2000)
7. Pearl, J.: Causality: Models, Reasoning, and Inference. Cambridge University Press, Cambridge (2009)
8. Bollen, K.A., Pearl, J.: Eight myths about causality and structural equation models. In: Morgan, S.L. (ed.) Handbook of Causal Analysis for Social Research, pp. 301–328. Springer, Dordrecht (2013). https://doi.org/10.1007/978-94-007-6094-3_15

9. Elwert, F.: Graphical causal models. In: Morgan, S.L. (ed.) Handbook of Causal Analysis for Social Research, pp. 245–273. Springer, Dordrecht (2013). https://doi.org/10.1007/978-94-007-6094-3_13
10. Menzies, P.: Causal models, token causation, and processes. Philos. Sci. **71**(5), 820–832 (2004)
11. Zholtkevych, G.N., Bespalov, Y.G., Nosov, K.V., Abhishek, M.: Discrete modeling of dynamics of zooplankton community at the different stages of an antropogeneous eutrophication. Acta. Biotheor. **61**(4), 449–465 (2013)
12. Kalman, R.E., Falb, P.L., Arbib, M.A.: Topics in Mathematical System Theory. McGraw-Hill, New York (1969)
13. Blalock, H.M.: Social Statistics. McGraw-Hill, New York (1972)
14. Boyd, D., Ellison, N.: Social network sites: definition, history, and scholarsh. J. Comput. Mediated Commun. **13**(1), 210–230 (2007)
15. Di Caprio, D., Santos-Arteaga, F.J., Tavana, M.: Self-organizing social networks by preference similarity and the networking capacity of their users. Procedia Comput. Sci. **83**, 496–503 (2016)
16. Chen, M., Mao, S., Liu, Y.: Big data: a survey. Mobile. Netw. Appl. **19**, 171–209 (2014)
17. Cole, T.J., Bellizzi, M.C., Flegal, K.M., Dietz, W.H.: Establishing a standard definition for child overweight and obesity worldwide: international survey. BMJ Clin. Res. **320**, 1–6 (2000)
18. ICES: Report of the Benchmark Workshop on Arctic Stocks (WKARCT), 26–30 January 2015. ICES Headquarters, Denmark. ICES CM 2015/ACOM:30 (2015)

ICT in Education

Implementation of an ICT Curriculum

Tips and Pitfalls of New Learning Concepts

Leo van Moergestel(✉), Ander de Keijzer, and Esther van der Stappen

HU University of Applied Sciences Utrecht, Utrecht, The Netherlands
leo.vanmoergestel@hu.nl
http://www.hu.nl

Abstract. This paper describes the process of redesigning a computer science (CS) curriculum and introducing blended learning in an CS educational program. The methodology that has been used as well as the motivation for the choices made are given. The first results compared with results from previous courses that used a more classical teaching approach are given. These results show that the new methodology proves to be promising and successful. The successes of the new program as well as the problems encountered are discussed. The solutions introduced in the second time the new courses were running are also presented.

Keywords: Blended learning · Course development · Didactic models in practice

1 Introduction

The rapid change in insights in how students learn and the technical possibilities has lead to innovative educational technologies. E-learning and Massive Open Online Courses (MOOCs) are examples of recent developments in Technology-Enhanced Learning (TEL). At the Utrecht University of Applied Sciences a project was started to design a new curriculum for bachelor students. There were several reasons for this project. The first reason was the fact that due to a reorganisation of several educational CS programs, there was a need for a new curriculum. The other reasons were new insights and the development of new concepts in blended learning as well as the strategic view of our university on future education. The combination of these reasons turned out to be a good starting point for the project.

One of the goals of the project was to construct this new curriculum using new concepts in higher education. Some of these concepts came into existence because of advances in information and computing technology. Nowadays all of our students have high speed access to the internet which opens the opportunity for video, and other multimedia applications that can be used by students at home in a so-called distant learning situation. By combining several kind of possibilities to learn for students a so-called blended learning environment comes

A. Ginige et al. (Eds.): ICTERI 2016, CCIS 783, pp. 117–132, 2017.
https://doi.org/10.1007/978-3-319-69965-3_7

into existence. This paper will discuss the approach that has been followed to use this blended learning approach in our course re-design in combination with the considerations and motivations for the choices made. A comparison with results from earlier courses is also given.

This paper is organised as follows: In the next section the concepts that are used will be explained. This is combined with a discussion of related work. Next the approach of curriculum re-design is presented in the section "Curriculum re-design". The section "Implementation" is dedicated to the implementation of the first part of the new bachelor program and the choices that have been made about teaching concepts. The first results are presented in the section "Results" followed by a discussion where the pitfalls and problems encountered are treated. The tips to avoid the pitfalls or solve the problems are also given. Finally a conclusion and a bibliography will end the paper.

2 Concepts and Related Work

The project is based on several innovative concepts in relation to education. In this section the concepts are introduced. In a later section an explanation will be given why these concepts have been used in our innovation project.

2.1 Blended Learning

Blended learning [6] was introduced as a mix between face-to-face and online learning, giving rise to the challenge of virtually limitless design possibilities.

The idea of blended learning is that the learning content and subjects should be offered to the student in several ways. The student can choose the type of explanation that fits herself the best. Also a combination of learning possibilities might help a student to understand the subjects. Whitelock and Jelfs [12] opened a journal special issue on this topic with three definitions:

- the integrated combination of traditional learning with web-based online approaches;
- the combination of media and tools employed in an e-learning environment; and
- the combination of a number of pedagogic approaches, irrespective of learning technology use.

Of these, the first is perhaps the most common interpretation. For our approach the third definition fits the best, but we also heavily rely on web-based approaches.

2.2 Flipping the Classroom

Flipping the classroom is an educational teaching method where lectures are replaced by self-study. This self-study is supported by moments of interaction between teacher and student where students can ask questions and put the theory

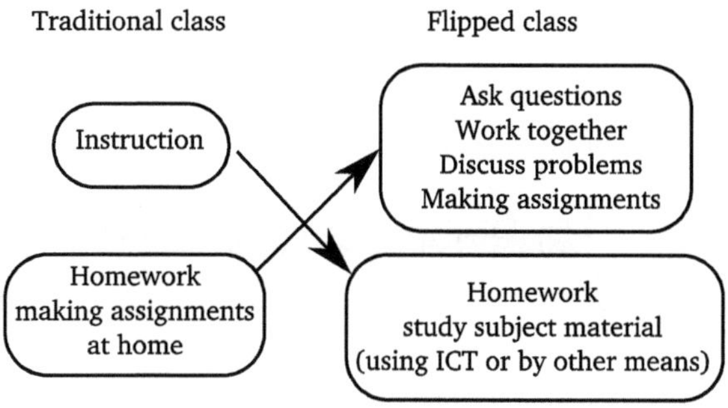

Fig. 1. Flipping the classroom

into practice. In Fig. 1 this situation is depicted. What is not shown in the figure is the fact that in the flipped class students should do their homework before the moment of interaction with the teacher and other students. The reason to start using this method is mostly based on the observation that the concentration of the student is limited to a small amount of time. Working on a problem related to theory that has been studied before gets the student more involved. A definition of flipping the classroom is given by Lage et al. [8]: *Inverting the classroom means that events that have traditionally taken place inside the classroom now take place outside the classroom and vice versa.* An overview of flipping the classroom is given by [1].

2.3 4C/ID-Model

The 4C/ID-model is a didactic model that is specially suited for training students that are determined to have a professional career in the industry [9]. 4C/ID is an abbreviation for 4 components instructional design. The main idea is that students will always work on real-life products that are closely related to their future professional work. In the starting phase of their training students are confronted with situations where most problems are already scoped and analysed by the teachers. Students have to fill in the gaps. Later on, students are given situations where they have to do more work themselves, while in the final part of their training they should be able to solve problems in complex situations by themselves. In all cases they are working on a real-life product fitting their educational program. The 4 components of the model are:

1. Learning task (backbone of the education program).
2. Supportive information.
3. Procedural information.
4. Part-task practice.

In Fig. 2 a schematic overview of the concept is shown. Students are working on learning tasks (1) represented by big circles in Fig. 2. At the beginning of their study, students start with simple tasks and students will work on complex tasks

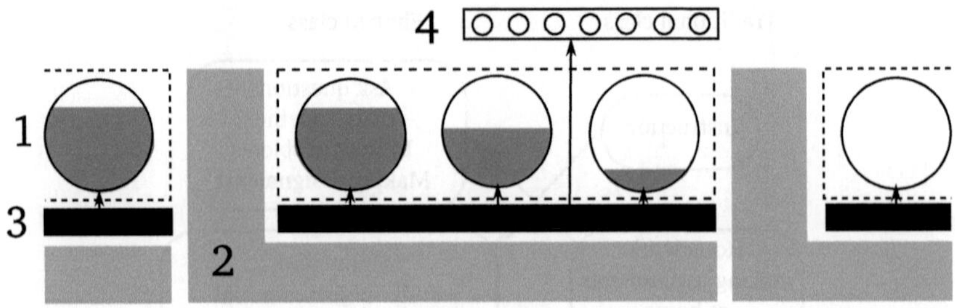

Fig. 2. The 4C/ID model [10]

at the end. The simplicity and complexity is controlled by giving much support in the beginning and reducing this support at the end. In Fig. 2 this is shown by the level the circles are "filled". The learnings tasks are part of a project (the dashed boxes, containing learning tasks) that will result in a product. The type of product the students will work on is related to their future professional career. Supportive information (2) is given at the beginning of every project. The supportive information gives the students a solid background to handle the problems they will encounter. This mostly theoretical information is also given during the time students are working on their tasks. The procedural information (3) is presented in a "just in time" manner to the students at the moment they really need it to complete the learning task. Practising with new concepts is also included in the model. In Fig. 2 this is represented by small circles (4). A thorough treatment of the concept can be found in [10].

2.4 Technology-Enhanced Learning

Two of the concepts mentioned in this section are closely related to IT. Flipping the classroom and blended learning heavily rely on IT and internet technology, while the 4C/ID model is a concept that can be implemented without any IT support.

3 Curriculum Re-design

Our main goal is to build a curriculum that uses new but proven concepts and fits in the educational concept that was chosen in the preliminary phase of the project. The concepts should be proven, because a situation where students are treated as guinea pigs in an experiment should in all cases be avoided. The quality and high standards used in the past should be guaranteed. The research presented in this paper has its main focus on the first part of this new curriculum. This first part consists of three courses of 5 European credits (EC's) [11] each. The courses are attended by around 500 first-year IT students.

3.1 Initial Situation

One year ago there were four different IT-based professional bachelor programs at the University of Applied Sciences Utrecht. These four programs were:

- Software engineering (SE). This program is meant for students who want to be a developer of complex software. Problem analysis, software design, implementation and testing are important aspects of this program.
- Computer science and embedded systems (CSES). CSES aims to ...
- Business IT and management (BIM). Students following this program should become the linking pin between business and IT. They should be able to analyse business processes and give advice how information technology can be used to improve these processes.
- System and network engineering (SNE). SNE is a program for students that has the focus on IT infrastructure design, implementation and maintenance. Network technology, security and infrastructure architecture are important aspects of this program.

These trajectories are now combined in one bachelor study with four different profiles. Combining bachelor studies was stimulated by the government for all existing bachelor studies. This will make it easier for students to select a certain study and specialize for a certain profile later. Combining these four given trajectories may seem like a simple reorganisation, but the students are offered more possibilities to select a set of courses leading to the final diploma. Cooperation between the different types of IT engineers is also more stimulated. This type of cooperation is often needed in the IT industry, so training students to cooperate during their study in multidisciplinary teams will support their professional career.

3.2 Design Principles

As a design principle we used the term MOS for the three properties every course should adhere to. MOS stands for:

- Motivation - use of new technology and didactical approaches should align more with students preferences. Working on real assignments (as in the 4C/ID model) and deeper discussions should motivate students more. Every course should be designed with the student in mind. It should motivate the student.
- Orientation - give students a good overview over the broad IT-landscape and professional roles therein. Guide them towards choosing one of the four specializations within the Bachelor program. After half a year a student should make a decision about which of the four tracks available she should follow.
- Selection - have the right level of difficulty in content and assignments so that passing 10 of the 12 courses more or less predicts finishing the study within the regular duration of four years.

Both orientation and selection are prescribed by regulations of the Dutch government [4].

3.3 Involvement of the Teaching Staff

At the beginning of the project, ideas and advice from the teaching team of around 40 persons were collected. Presentations and discussion sessions were organised. During two weeks discussion posters were available where all people involved could post their comments and remarks. The whole teaching and management staff was able to contribute ideas and proposals. This resulted in the following set of requirements:

- The same program for all students during the first half year.
- No instruction to big groups, but classes with a maximum of 32 students.
- Showcases of real-life products as the main drive for the curriculum.
- Special attention to the development of professional skills by the students during their training. Professional skills are skills that one might expect from a professional IT engineer apart from the specific domain related knowledge and skills, such as professional written and oral communication, leadership, (team) planning and ethical skills.

3.4 Involvement of Stakeholders

From the beginning the IT industry as well as the students were involved. The plans for the curriculum were presented to representatives of the IT industry and students from all years of the four-year curriculum. By using the feedback of these meetings, the new curriculum adheres to the requirements and expectations of the software industry as well as the students. Every course under development was also checked this way.

The developers of different courses had several meetings to guarantee the coherence between the different courses that were given in parallel. All courses started with an explanation of this coherence and why this specific course had its place in the bachelor program.

4 Curriculum Fundamentals

To implement a curriculum for bachelor degree students the first thing that should be done is to investigate and determine the final level of education. What capabilities should a student have at the end of the educational program? For this aspect two documents played an important role. The European Competence Framework (e-CF) [3] and the domain description of the bachelor of ICT issued by the cooperating ICT institutes of the Dutch universities of applied sciences HBO-i ((Hoger Beroeps Onderwijs informatica), Higher Professional Education ICT) [7]. We will give a short description of both documents.

4.1 e-CF and EQF

In 2001, a number of large ICT companies expressed their concern about the shortage of qualified ICT professionals on the job market. The e-CF is developed for businesses and human resource management, and it uses levels of proficiency for the whole of the ICT professional field. Many jobs in the ICT industry are described together with the level of competence required. As stated in the document [3]: *reference framework of competences applied within the Information and Communication Technology (ICT) sector that can be used and understood by ICT user and supply companies, ICT practitioners, managers and Human Resources (HR) departments, the public sector, educational and social partners across Europe.*

Apart from e-CF there is also the European Quality Framework (EQF) [5] that is a uniform framework to make qualifications more comparable across Europe. It defines 8 levels and for each level there is a description for knowledge, understanding and skills. Level 6 is the bachelor level.

4.2 HBO-i

The model proposed by HBO-i is actually a matrix of competences based on lifecycle activities and architectural layers. The lifecycle activities are derived from the lifecycle of information systems. These activities are: management, analyse, advice, design and implementation of information systems. The architectural layers beginning with the starting point from the ICT perspective is hardware interfacing, on top of this layer is software, resulting in the layer infrastructure. On top of the infrastructure there is the business layer and finally we have the user interaction layer. This results in a matrix as depicted in Fig. 3. For every field in this matrix, three skill levels are defined and described. These levels are determined by the degree of autonomy, the behaviour expected (from being responsible for a single task, to leadership or control of a project) and the complexity of the context (from predictable and stable to unpredictable). These skill levels are also related to the e-CF and EQF from the previous section. As an

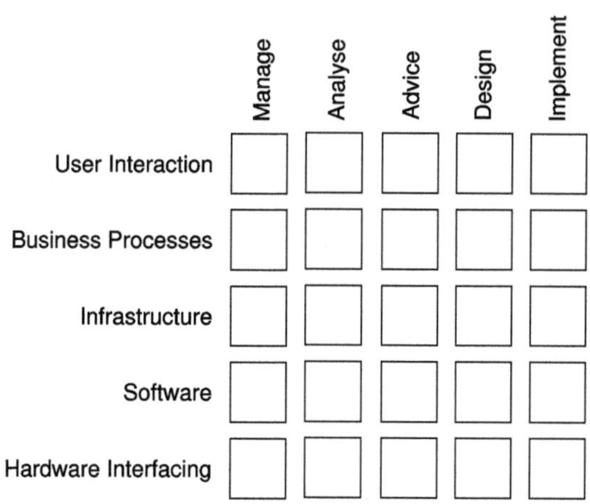

Fig. 3. Competence matrix

example, short descriptions of the skill levels for implementation at the software level are:

1. *Build and test an elementary software system and make it available.*
2. *Build and make available a software system consisting of multiple subsystems, and using existing components.*
   Integrate software components within an existing system, whilst monitoring the integrity and system performance. Perform regression tests.
   Perform and report on unit, integration and system tests.
3. *Build and make available a software system in line with existing systems and on the basis of the designed architecture, using existing frameworks.*
   Using test automation when performing tests.

An elaborate treatment and discussion of all levels for all cells in the matrix can be found in the publication by HBO-i [7].

4.3 End Levels

The fundamentals of our courses were based on the HBO-i matrix biased by the e-CF. These two concepts are well documented and also widely supported by universities of applied sciences and industries. Our university is also involved with and participating in the developments guided by HBO-i.

It depends on the type of ICT professional, what field at what levels should be reached at the end of the 4 year course. For the different specialisations the required levels at the end of the bachelor training were defined. The e-CF played a role in this decision. The result was also discussed with professionals from the industry to prevent blind spots and to get a clear interpretation of what the industry expected as the bachelor end levels. As an example, in Fig. 4 the levels for software engineering are shown. So for every specialisation it was now clear in what end level the complete set of courses during the years should result. All specialisations should have at least 2 skills at level 3, 4 skills at level 2 and 7 skills at level 1. This way al courses should have the same complexity, be it within a different domain.

4.4 Increase of Levels per Year

In the previous section, the end levels were introduced. To reach these levels, a growing path was defined. At the end of the first year, 11 skills at level 1 should be reached. In Fig. 5 this matrix is shown for software engineering. The nice thing about this approach is that the development team for the courses also had a clear goal in mind. For these courses the levels were used as a starting point as well as the way the achievement of these levels were to be proven by the students involved.

The increase of levels has also been determined for the other years of the bachelor level. This results in increments of the levels that should be implemented in the courses involved.

	Manage	Analyse	Advice	Design	Implement
User Interaction	1	1	1	2	2
Business Processes		1			
Infrastructure		1	1		
Software	2	2	1	3	3
Hardware Interfacing					

Fig. 4. End level competence matrix for software engineering

	Manage	Analyse	Advice	Design	Implement
User Interaction	1	1		1	1
Business Processes		1			
Infrastructure		1	1		
Software	1	1		1	1
Hardware Interfacing					

Fig. 5. Competence matrix for software engineering first year

4.5 Professional Skills

Apart from to the competences defined by the HBO-i matrix, attention should also be paid to the development of professional skills. By these skills are meant the competences a bachelor of ICT should have to act as a professional in a business environment. Examples of these skills are, the ability to do research, to give a presentation to an audience of peers, to acquire new knowledge and insight in new developments, cooperating in groups etc.

A list of skills required was set up. In the courses and projects the developers stated what professional skills were developed by the students taking the course of doing the project. Attention was paid to completeness of skills developed by the students.

5 Implementation

For the time being, only the first year will be considered. The first year curriculum consists of four periods of ten weeks. At the end of the year every student should have a personal assessment to check her knowledge and practical capabilities learned so far.

5.1 Learning Model

For instruction a selection of one the following possibilities has to be made:

1. Blended learning: use all kind of teaching techniques to train the student. Let the student decide which one fits her the best.
2. Problem driven education: students are confronted with a problem en should discover by themselves what knowledge and information they need to sole the problem. When they have questions, the teacher will get involved.
3. Project based education: a project is the central part of the training and should control all other educational methods involved.
4. Classical approach, giving theoretical training supported by practising the learned material.

Considering the new possibilities in IT, the blended learning option has been chosen. By definition, all other methods could be included, but the problem driven possibility had been used is the past and turned out to be not so successful. A pure classical approach is not apt for our education, because the type of student in our institute is more interested in practical problem solutions, than pure academic knowledge. Pure project based education that has been used in the past, seems to be missing some essential aspects. It turned out to be difficult to find a set of projects that covers all end-goals of a professional IT engineer. A choice has been made to combine blended learning with flipping the classroom. This approach did not need the instruction sessions for a huge number of students, that turned out to be not so successful in the past.

Next a choice has to be made between 4C/ID and the classical approach. In our case the 4C/ID model has been chosen because our institute focusses on training bachelor students to act as a professional worker in the IT industry. The teaching team had the experience that these students are more motivated when they are working on real products [9]. This in contrast with the pure academic approach where the student is trained to be a scientist. In the academic approach the coherence with the work in the IT industry is less evident. The 4C/ID also fits well in the concept of blended learning and flipping the classroom. Students are working on projects and will also be instructed to guarantee their theoretical knowledge.

The choices made also fit well with the design principles regarding motivation as explained in Sect. 3.2.

For the first year, four types of projects were developed. The first project was actually a set of three small projects, related to the three theoretical courses, that were mostly meant to orientate and motivate the students. In the second type of project, just before the students will select their specialisation, a so called intradisciplinary project had to be carried out. In this project with a higher complexity as the previous ones, a student should play a role that fits the specialisation chosen by that student. For each of the specialisations a role is defined in the project. This way a student can experience the role she has chosen and perhaps decide to switch to another specialisation if choosing this role turn out to be a complete mistake. The third type of project is also group based like the previous project, but here the specialisation has been chosen and within the project a group within the same specialism is working on a rather complex task. Finally, during the last project, the student has to work on a project on her own to prove that the student is capable to complete a complex task without the support of other team members.

5.2 Selection of Courses

To give the students the possibility to orient on the possibilities in a later phase of the bachelor program, all three courses should more or less be tied to the final four tracks the student has to select during the first half year. Table 1 shows the relationship between the courses and the four tracks. The orientation on the tracks is now guaranteed. A mark means that the course has a significant relation with the track mentioned. A missing mark does not mean that the course is not significant for the track, but that it is not a main focus. Another reason why these three courses are combined in the first period of the study is that it is a nice way to demonstrate the layered structure that plays an important role in many IT concepts. At the top is IT in organisations, The middle layer is presented by programming and finally the bottom layer (hardware, operating systems and networking) is covered by computer systems and networks. Having a model and three courses to start with, we will present the results in the next section.

6 Results

The new program started in September 2015 and the first period consisted of the three aforementioned courses. These courses had the same learning goals as

Table 1. Relationship between track and course

Education	Computers	Programming	Organisation
BIM			✓✓
SE		✓	✓
SNE	✓		✓
CSES	✓	✓	

courses we had in our old model so a comparison with courses in the previous year is legitimate. Some minor differences exists that will be discussed later. As mentioned and motivated in the previous section the courses are:

- introduction in programming
- computer systems and networks
- IT in organisations

In Table 2 the results for the three courses are given for both the blended and classical learning (previous year). Nb is the number of students doing the final exam for the blended course, Nc is the number of students doing exam for the previous classical course, pass is the number of student that passed the exam (including a second chance exam) and avg. is the average score for the exam on a scale from 1 to 10. The average score has not significantly changed, while the percentage of students passing the exam has increased.

Table 2. Results

Course	Nb.	Pass	Avg.	Nc.	Pass	Avg.
IT in organisations	474	372	5,7	413	247	5,7
Programming	486	399	6,1	425	330	6,2
Computer systems & networks	490	351	5,4	454	293	5,3

Figure 6 shows the results of the percentage of students that successfully completed the course in the regular time after a second chance had been offered. In our institute, every course will have a second exam every year. In the same figure, the results for the previous year have been added. From the figure it becomes clear that the yield has significantly increased for all courses. The most successful course so far was programming, though this course was also the most successful one in the previous year. There is also a slight difference in the content of the program, mainly due to the fact that Python has been used instead of PHP or Java may have influenced the outcome. The learning goals did not change, so it is fair to state that the new blended course was more successful than the year before. All other courses show a better outcome than the year before. The other two courses are quite similar in content and learning goals, so the results can be compared. One should keep in mind that the new curriculum was not only based on blended learning and the 4C/ID model, but also based on the fact that we did not separate the different specialisations from the start. Also the fact that there were no classes with more than 32 students should be mentioned here. In the previous year there were instruction classes for over 100 students. So the positive effect should be considered as the consequence of all changes in the curriculum.

7 Discussion

In this section, some of the problems encountered during development and introduction of the new curriculum are discussed as well as the successes achieved.

7.1 Pitfalls

Though the concepts presented were supported by the majority of teachers in our institute, some remarks should be made here:

1. The amount of work for the developers of the courses was much more than expected in the beginning. Selecting good online material took much more time, because the developers should check the quality and adequateness carefully [2]. Also the match of the online material with the learning goals turned out to be cumbersome. The alternative would have been developing material at our institute, but that would have cost even more time.
2. The flipping the classroom concept only works well if there is a possibility to monitor the self-study phase of the students. Students should study in advance and do some self tests. The results of these tests will give a clue

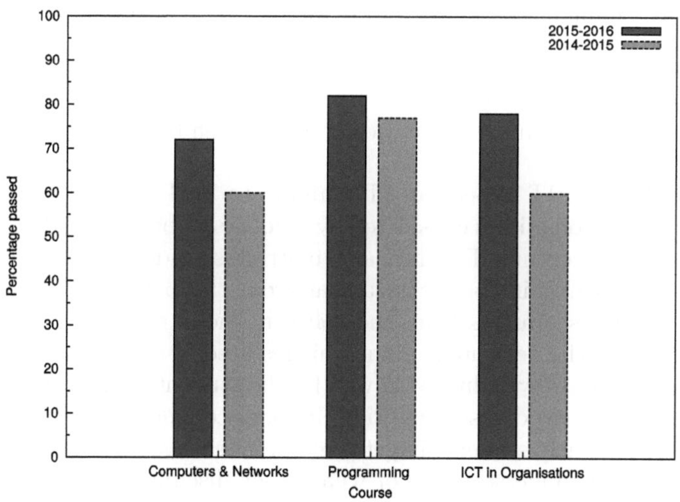

Fig. 6. Success rate per course

to the teacher which parts of the teaching material should be treated at the beginning of the classroom meeting before students will work on instructions and tasks related to the material. In our situation monitoring was not yet possible, so we had to ask the students what problems they encountered. This might lead to the situation that a group of students decides to pay less attention to the self-study, knowing that the teacher cannot check and almost certainly will explain the difficult parts at the beginning of the meeting.
3. There was quite a big difference in applying the concepts as proposed by different persons teaching the same course. Many more meetings would have helped to solve this problem, but this would also put a burden on the people involved.
4. Students were not used to the new educational concepts. More effort should be taken to introduce students to the new environment as well as explaining the system.

The problems mentioned here are solvable. In the situation described under number 1 and 3 it means making more time available for the teaching staff to prepare the courses and having the possibility to discuss the method and help each other to use good practices. Currently the solution for the problem described under number 2 is under development. Monitoring activities per individual student is necessary and as an Institute of ICT we are strongly involved with the software used to create the blended learning environment. The problem described under number 4 should be solved by explaining the students the concept more carefully. The fact that students know that their home-activities for the course are monitored may also help to activate the students.

7.2 Successes

There is also a number of successes to report on the new program.

1. The new learning paradigm turned out to help a bigger number of students to complete the courses within due time. Given the fact that some important improvements still can be made, we expect an even better result for the years to come.
2. Most students in IT expect an IT training to be IT-based. Older ways of instruction are sometimes considered by students to be outdated or belonging to the previous century. This fact also motivates students.
3. Another positive point is the fact that from the beginning both students as well as representatives from the software industry were involved in the construction of the new curriculum. This resulted in a curriculum that was a wide support both by industry as well by the student community.
4. A feature of the new course is multidisciplinary cooperation between students in project. This help students in getting a broader view on the IT domain and getting a better understanding of what other disciplines in that field mean and accomplish.

5. In the new situation students will select a specific specialisation after half a year of their study. This makes it much easier for students to find out what specialisation fits the best with their interest and capabilities. In the older setup there was also a half year with almost similar programs, but the switch to another discipline was more complicated. Actually the student had to switch while in the current course she only decides what specialisation she will choose after half a year.

Given the fact that the pitfalls we encountered are solvable, the switchover to the new educational approach can be considered as a success so far.

8 Conclusion and Discussion

In this paper we described a project to renew a CS curriculum. In this curriculum concepts like 4C/ID, blended learning and flipping the classroom play an important role. The concepts that have been used are introduced and explained as well as the implementation and first results. The first results are promising.

IT-based solutions offer great opportunities for educational renewal, but care has to be taken about the way they are used. Also the time and work involved was in our case underestimated leading to a result that is not yet at the level that we had in mind at the start. However, as mentioned before the results so far are promising and give support to the idea that this is a good approach to keep students involved and motivated with their study at the educational institute.

References

1. Bishop, J., Verleger, M.: The flipped classroom: a survey of the research. In: Proceedings of the American Society for Engineering Education (ASEE) Annual Conference (2013)
2. Chauhan, J., Goel, A.: An analysis of video lecture in MOOC. In: Proceedings of the 11th International Conference on ICT in Education, Research and Industrial Applications (ICTERI), Lviv, pp. 35–50 (2015)
3. European e-Competence Framework 3.0: A common European Framework for ICT Professionals in all industry sectors (2014). http://www.ecompetences.eu
4. Dutch: Wet kwaliteit in verscheidenheid hoger onderwijs (dutch). Staatsblad van het Koninkrijk der Nederlanden (Dutch), vol. 298, pp. 1–23 (2013)
5. European Commission: The European qualifications framework for lifelong learning (EQF) (2008)
6. Garrison, D.R., Kanuka, H.: Blended learning: uncovering its transformative potential in higher education. Internet High. Educ. **7**(2), 95–105 (2004)
7. HBO-i: Domain description Bachelor of ICT (2014). http://www.hbo-i.nl/boict-english/
8. Lage, L., Platt, G.: The internet and the inverted classroom. J. Econ. Educ. **31**, 11 (2000)
9. Merriënboer, J., Clark, R., de Croock, M.: Blueprints for complex learning: the 4C/ID-model. ETR&D **50**(2), 39–64 (2002)

10. Merriënboer, J., Kirschner, P.: Ten Steps to Complex Learning. Lawrence Erlbaum Associates Inc. (2007)
11. Wagenaar, R.: An introduction to the European credit transfer and accumulation system (ECTS). EUA, Bologna Handbook (2006)
12. Whitelock, D., Jelfs, A.: Special issue on blended learning. J. Educ. Med. **28**(2–3), 99–100 (2003)

Robotics and Prosthetics at Cleveland State University: Modern Information, Communication, and Modeling Technologies

Yuriy Kondratenko[1,2(✉)], Gholamreza Khademi[1], Vahid Azimi[1], Donald Ebeigbe[1], Mohamed Abdelhady[1], Seyed Abolfazl Fakoorian[1], Taylor Barto[1], Arash Roshanineshat[1], Igor Atamanyuk[2], and Dan Simon[1]

[1] Department of Electrical Engineering and Computer Science, Cleveland State University, Cleveland, OH, USA
y_kondrat2002@yahoo.com, d.j.simon@csuohio.edu
[2] Department of Intelligent Information Systems, Petro Mohyla Black Sea National University, Mykolaiv, Ukraine

Abstract. This chapter concentrates on the correlation between research-based education, government priorities and research funding. Special attention is paid to an analysis of the role of modern information and communication technology (ICT) in the education of engineering students. Successful cases with specific description of computer modeling methods for the implementation of prosthesis and robotics research projects are presented based on experiences in the Embedded Control Systems Research Laboratory of Cleveland State University.

Keywords: Robotics · Prosthetics · Modeling · Research-based education

1 Introduction

Information and communication technologies (ICT), mathematical modeling, and computer simulation play a significant role in higher education. Most advanced educational systems in the world are oriented toward the implementation of educational processes of modern ICT and software for modelling and simulation in various fields of human activity, including science, engineering, and technology. This approach is required for the efficient training of students at various levels: undergraduates, graduates, and doctoral students. Many international conferences on ICT and its applications for education are devoted to the use of computer modeling, open-source software, pedagogical e-learning, web-based e-learning, course-centered knowledge management and application in online learning based on web ontology, on-online learning in enterprise education, simulation languages, modeling and simulation for education and training, improving education through data mining, 3D software systems, 3D visualization, wireless communication, experimental teaching of program design, different approaches in teaching programming, web-based computer-assisted language learning, and so on.

A. Ginige et al. (Eds.): ICTERI 2016, CCIS 783, pp. 133–155, 2017.
https://doi.org/10.1007/978-3-319-69965-3_8

It is important that university and IT-industry participants of conferences try to find efficient solutions for the abovementioned computer-modeling-based educational problems. For example, participants from 178 different academic institutions, including many from the top 50 world-ranked institutions, and from many leading IT corporations, including Microsoft, Google, Oracle, Amazon, Yahoo, Samsung, IBM, Apple, and others, attended the 12th International Conference on Modeling, Simulation and Visualization Methods, MSV-2015, in Las Vegas, Nevada, USA.

If IT industry today supports higher education, then tomorrow's IT-based companies, government research agencies, and national laboratories will obtain the high-quality graduates that they need. New achievements in ICT require continuous tracking by educators, and implementation in education.

Successful introduction of ICT to higher education based on research-oriented education and training is considered and analyzed in this chapter. The focus is on the role of computer modeling and simulation in prosthesis and robotics research for increasing student quality, including grading their practical skills, and including efficient professor-student interactions.

This chapter is organized as follows. Section 2 reviews related literature and discusses the challenge of integrating research and education. Section 3 summarizes research-based education at Cleveland State University (CSU). Section 4 summarizes seven research-based education projects at CSU. Section 5 briefly discusses the common training, skills, and educational program at CSU that enables the success of research-based education. Section 6 concludes the chapter.

2 Related Works and Problem Statement

Many publications are devoted to teaching methods and approaches based on ICT and computer modelling, for increasing the efficiency of their interrelation: qualitative modeling in education [3], computer simulation technologies and their effect on learning [24], opportunities and challenges for computer modeling and simulation in science education [34], web-based curricula [4] and remote access laboratories, computer-based programming environments as modelling tools in education and the peculiarities of textual and graphical programming languages [17], interrelations between computer modeling tools, expert models, and modeling processes [44], efficient science education based on models and modelling [9], educational software for collective thinking and testing hypotheses in computer science [26], and others.

A lot of publications deal with improving teaching efficiency for specific courses by introducing modern ICT and computer modelling technologies. In particular, modelling supported course programs, computer-based modelling (AutoCAD, Excel, VBA, etc.) and computer system support for higher education in engineering [8]; software to enhance power engineering education [35]; computer modeling for enhancing instruction in electric machinery [23]; computer modelling in mathematics education [40]; GUI-based computer modelling and design platforms to promote interactive learning in fiber optic communications [45]; RP-aided computer modelling for architectural education [36]; teaching environmental modelling; computer modelling and simulation in power electronics education [27]; and a virtual laboratory for a communication and computer networking course [22].

Special attention in the literature [5, 15, 16] is paid to the role of ICT and modeling technology in education and training in the framework of research-based curricula. This educational approach deals first with educational directions such as robotics, mechatronics, and biomechanics (RMBM) [12, 33, 41]. The correlation of RMBM with ICT and modeling are underlined by results such as: a multidisciplinary model for robotics in engineering education; integration of mechatronics design into the teaching of modeling; modelling of physical systems for the design and control of mechatronic systems [41]; biomechanical applications of computers in engineering education [33]; computerized bio-skills system for surgical skills training in knee replacement [6]; computer modelling and simulation of human movement [25]; computer modelling of the human hand [19]; and design and control of a prosthesis test robot [29, 30].

This chapter builds upon, and extends, the references discussed above. The basic classroom teaching methods, approaches, and specific courses at CSU are similar to those at universities across the world. However, those characteristics are not the primary determinants of research-based education. This chapter presents the features of research-based education at CSU by reviewing seven specific graduate student-led research projects. As the research projects are discussed in the following sections, the reader will note their commonalities, including common tools, research approaches, motivation, and societal focus. The main aims of this chapter are given as follows.

(a) Description and analysis of research-based education based on the experience in the Embedded Control Systems Research Laboratory at the Electrical Engineering and Computer Science Department at the Washkewicz College of Engineering at Cleveland State University (CSU), USA, with a focus on undergraduate, graduate, and doctoral student participation in prosthesis and robotics research, which is funded by the US National Science Foundation (NSF);
(b) Analysis of applied ICT and modeling technologies and advanced software, as well as their implementation in student research, including course work, diploma projects, and Doctoral, Master's, and Bachelor's theses;
(c) Focus on the correlation between student research and government science priorities based on successful cases of ICT and advanced modelling implementation in US government-funded prosthesis research, with particular focus on undergraduate, graduate, and doctoral student participation in prosthesis and robotics research.

3 Research-Based Education and Government Priority Project

CSU's research project "Optimal prosthesis design with energy regeneration" (OPDER) is funded by the US NSF (1.5 M USD). Professors and students from the Department of Electrical Engineering and Computer Science, and the Department of Mechanical Engineering, are involved in research according to the project goals, which deal with the development of: (a) new approaches for the simulation of human limb control; (b) new approaches for optimizing prosthetic limb control, capturing energy during walking, and storing that energy to lengthen useful prosthesis life; (c) prosthesis prototype development.

The human leg transfers energy between the knee, which absorbs energy, and the ankle, which produces energy. The prosthesis that results from this research will mimic the energy transfer of the human leg. Current prostheses do not restore normal gait, and this contributes to degenerative joint disease in amputees. This research will develop new design approaches that will allow prostheses to perform more robustly, closer to natural human gait, and last longer between battery charges.

This project forms a framework for research-based education. Doctoral, graduate, and undergraduate students are involved in research such as: the study of able-bodied gait and amputee gait; the development of models for human motion control to provide a foundation for artificial limb control; the development of electronic prosthesis controls; the development of new approaches for optimizing prosthesis design parameters based on computer intelligence; the fabrication of a prosthesis prototype and its test in a robotic system; the conduct of human trials of the prosthesis prototype.

The role of student participation in all aspects of the research is significant for increasing their qualifications for their careers, for presentations at conferences, for publishing in journals, and for research with professors who can help them be more successful in building their future careers in industry or academia. In the next section we describe the student contribution to prosthesis and robotics research at CSU.

4 Student Contributions to Prosthesis and Robotics Research

Successful cases of student research in the framework of the OPDER project are described in this section.

Evolutionary Optimization of User Intent Recognition (UIR) for Transfemoral Amputees. Powered prostheses are being developed to help amputees handle several different activities: standing, level walking, stepping up and down, walking up and down a ramp, etc. For each walking mode, a different control policy is required to control the prosthesis. User intent recognition system plays an important role to infer the user's activity mode while transitioning from one walking mode to another one, and then to activate the appropriate controller. Pattern recognition techniques are used to address such problems.

In this research, mechanical sensor signals are experimentally collected from an able-bodied subject, and comprise the training inputs to the UIR system. Signals are processed and filtered to eliminate noise and to handle missing data points. Signals reflecting the state of the prosthesis, user-prosthesis interactions, and prosthesis-environment interactions are used for user intent recognition. Hip and ankle angles, ground reaction force (GRF) along three axes, and hip moment are chosen as relevant input signals that reflect various gait modes. Principal component analysis is used to convert data to a lower dimension by eliminating the least relevant features. We propose the use of correlation analysis to remove highly correlated observations from the training set.

The main component of the UIR system is its classifier. We use K-nearest neighbor (KNN) as a classification method for this purpose. KNN is modified and optimized with an evolutionary algorithm for enhanced performance. We also modify KNN so that the contribution of each neighbor is weighted on the basis of its distance to the test point, and on the basis of the history of previously classified test points. This modification leads to better performance than standard KNN. Optimization techniques can be used to tune the KNN parameters and obtain a classification system with the highest possible accuracy. We choose biogeography-based optimization (BBO) as the evolutionary optimization algorithm for this purpose. The optimization problem is to minimize the classification error. The UIR system can then be used to identify unknown walking activities. The architecture of the UIR system is illustrated in Fig. 1. We use MATLAB to implement user intent recognition. BBO is a stochastic algorithm, so it requires several runs to optimize the parameters. We use parallel computing to reduce the optimization time from 7.8 days to about 20 h [11]. To test the proposed method, multiple sets of experimental data are collected for various gait modes: standing (ST), slow walking (SW), normal walking (NW), and fast walking (FW). Figure 2 illustrates the experimental setup for able-bodied subjects. Future work will extend the proposed approach to amputee gait data. Figure 3 shows an example of test data for a walking trial lasting approximately 18 s, which includes different walking modes.

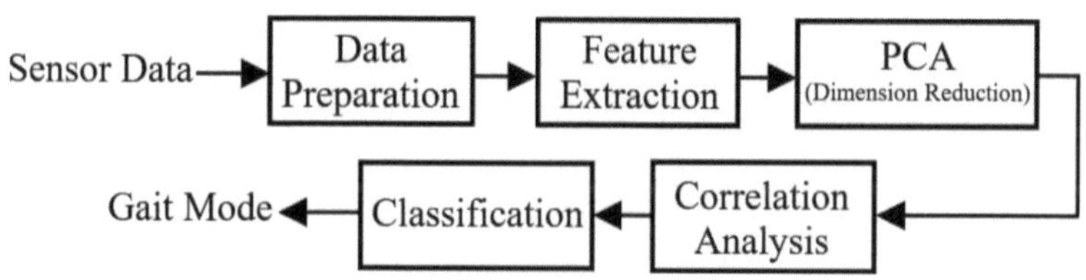

Fig. 1. Architecture of user intent recognition system: an evolutionary algorithm (not shown) is used to optimize the system components

Fig. 2. Experimental setup: data collection for user intent recognition for able-bodied subjects

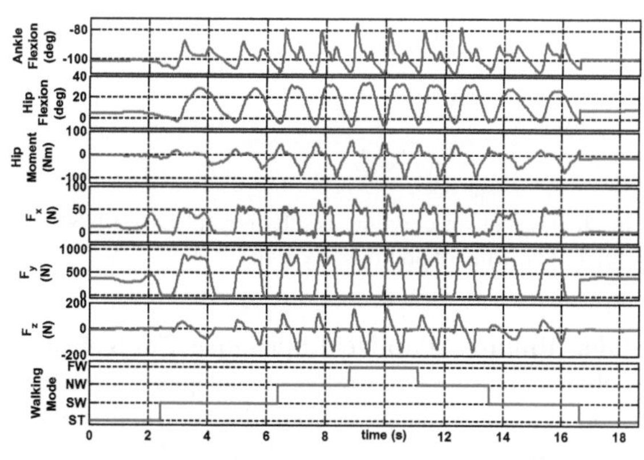

Fig. 3. Sample test data showing four different gait modes and transitions: ST (standing), SW (slow walk), NW (normal walk), and FW (fast walk)

Table 1 shows the performance of different versions of KNN. The first row of Table 1 shows simple KNN, which uses $K = 7$ nearest neighbors and results in 12.9% test error. Test error reduces to 11.5% if we use weighted KNN with $K = 7$ nearest neighbors. Test error is 8.06% when weighted KNN is used in addition to previously classified gait modes to inform the current classification mode. The fourth row of Table 1 shows that the optimized weighted KNN with information from previously classified modes provides the minimum classification error. Figure 4 shows the performance of the classifier using both simple KNN and optimized KNN. Classification error for optimized KNN is 3.59% compared to 12.9% with standard KNN.

In conclusion, KNN was modified to enhance the performance of a user intent recognition system. An evolutionary algorithm was applied to optimize the classifier parameters. Experimental data was used for training and testing the system. It is shown that the optimized system can classify four different walking modes with an accuracy of 96%. The code used to generate these results is available at http://embeddedlab.csuohio.edu/prosthetics/research/user-intent-recognition.html. Further details about this research can be found in [11].

Table 1. Performance of KNN for user intent recognition with different classifier parameters

Method	Train error	Test error
Simple KNN ($K = 7$)	7.41%	12.9%
Weighted KNN ($K = 7$)	3.81%	11.5%
Weighted KNN and Recent Modes ($K = 7$)	3.44%	8.06%
Optimized KNN ($K = 12$)	3.22%	3.59%

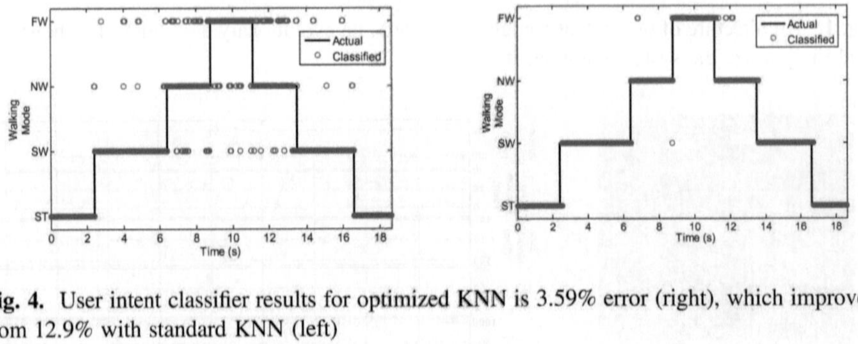

Fig. 4. User intent classifier results for optimized KNN is 3.59% error (right), which improved from 12.9% with standard KNN (left)

Stable Robust Adaptive Impedance Control of a Prosthetic Leg. We propose a nonlinear robust model reference adaptive impedance controller for a prosthesis test robot. We use an adaptive control term to compensate for the uncertain parameters of the system, and a robust control term to keep the error trajectories in a boundary layer so the system exhibits robustness to variations of ground reaction force. The algorithm not only compromises between control chattering and tracking, but also limits tracking-error-based (TEB) parameter adaptation to prevent unfavorable drift. The acceleration-free regressor form of the system obviates the need to measure joint accelerations, which would otherwise introduce noise in the system. We use particle swarm optimization (PSO) to optimize the parameters of the controller and the adaptation law. The PSO cost function is comprised of torque optimality and tracking performance.

The prosthesis is an active transfemoral (above-knee) prosthesis. The complete system model has a prismatic-revolute-revolute (PRR) joint structure. Human hip and thigh motion are emulated by a prosthesis test robot. The vertical degree of freedom represents vertical hip motion, the first rotational axis represents angular thigh motion, and the second rotational axis represents prosthetic angular knee motion [1, 2]. The three degree-of-freedom model can be written as follows:

$$M\ddot{q} + C\dot{q} + g + R = u - T_e, \tag{1}$$

where $q^T = [q_1 \quad q_2 \quad q_3]$ is the vector of generalized joint displacements (q_1 is the vertical displacement, q_2 is the thigh angle, and q_3 is the knee angle); u is the control signal that comprises the active control force at the hip and the active control torques at the thigh and knee; and T_e is the effect of the GRF on the three joints. The contribution of this research is a nonlinear robust adaptive impedance controller using a boundary layer and a sliding surface to track reference inputs in the presence of parameter uncertainties. We desire the closed-loop system to provide near-normal gait for amputees. Therefore, we define a target impedance model with characteristics that are similar to those of able-bodied walking:

$$M_r(\ddot{q}_r - \ddot{q}_d) + B_r(\dot{q}_r - \dot{q}_d) + K_r(q_r - q_d) = -T_e \tag{2}$$

where q_r and q_d are the trajectory of the reference model and the desired trajectory respectively. Since the parameters of the system are unknown, we use the control law

$$u = \widehat{T}_e - K_d \,\mathrm{sat}(s/\widehat{\mathrm{diag}}(\varphi)) + \widehat{M}\dot{v} + \widehat{C}v + \widehat{g} + \widehat{R} \tag{3}$$

where the diagonal elements of φ are the widths of the saturation function; $s = \dot{e} + \lambda e$ and $v = \dot{q}_r - \lambda e$ are the error and signal vectors respectively; $e = q - q_r$ denotes tracking error, $\lambda = \mathrm{diag}(\lambda_1, \lambda_2, \ldots, \lambda_n)$, where $\lambda_i > 0$; $K_d = \mathrm{diag}(K_{d1}, K_{d2}, \ldots, K_{dn})$, where $K_{di} > 0$; n is the number of rigid links; and $\widehat{M}, \widehat{C}, \widehat{g}, \widehat{R}$, and $\widehat{T}_e$ are estimates of M, C, g, R, and T_e respectively. The control law [36] of Eq. (3) comprises two different parts. The first part, $\widehat{T}_e - K_d \mathrm{sat}(s/\mathrm{diag}(\varphi))$, satisfies the reaching condition ($\mathrm{sgn}(s)\dot{s} \leq -\gamma$, $\gamma = [\gamma_1 \quad \gamma_2 \quad \ldots \quad \gamma_n]^T$ and $\gamma_i > 0$) and handles the variations of the external inputs T_e. The second part, $\widehat{M}\dot{v} + \widehat{C}v + \widehat{g} + \widehat{R}$, is an adaptive term that handles the uncertain parameters, which are estimated via the following adaptation mechanism:

$$\dot{\hat{p}} = -\mu^{-1} Y^T(q, \dot{q}, v, \dot{v}) s_\Delta, \tag{4}$$

where $Y(q, \dot{q}, v, \dot{v})$ is an acceleration-free regressor for the left side of Eq. (1); s_Δ is the boundary layer trajectory; and μ is an $r \times r$ design matrix with positive diagonal elements. To trade off control chattering and tracking accuracy, and to create an adaptation dead zone to prevent unfavorable parameter drift, we define a trajectory s_Δ as follows [1, 2, 39]:

$$s_\Delta = \begin{cases} 0, & |s| \leq \operatorname{diag}(\varphi) \\ s - \varphi \operatorname{sat}(s/\operatorname{diag}(\varphi)), & |s| > \operatorname{diag}(\varphi) \end{cases} \tag{5}$$

where s_Δ is an n-element vector; the region $|s| \leq \operatorname{diag}(\varphi)$ is the boundary layer and the inequality is interpreted element-wise; and the diagonal elements of φ are the boundary layer thicknesses and the widths of the saturation function so that $\varphi = \operatorname{diag}(\varphi_1, \varphi_2, \ldots, \varphi_n)$ and $\varphi_i > 0$.

To perform a stability analysis of the controller, the following positive-definite Lyapunov function is considered:

$$V(s_\Delta, \tilde{p}) = \frac{1}{2}\left(s_\Delta^T M s_\Delta\right) + \frac{1}{2}\left(\tilde{p}^T \mu \tilde{p}\right). \tag{6}$$

Let us assume that $\left|\widehat{T}_{e_i} - T_{e_i}\right| \leq F_i \leq F_m$, $\gamma_m = \max(\gamma_i)$, and a is a positive scalar. Given the Lyapunov function of Eq. (6), the control law of Eq. (3), and the TEB adaptation mechanism of Eq. (4) in conjunction with the boundary layer trajectory of Eq. (5), if $K_{di} \geq -a\dot{q}_{max}\varphi_i + F_m + \gamma_m$, then $\dot{V}(s_\Delta, \tilde{p}) \to 0$ as $t \to \infty$, which means that $s_\Delta \to 0$ and the controller guarantees the convergence of the error trajectories to the boundary layer after the adaptation period.

We use PSO to tune the controller and estimator parameters. PSO decreases the cost function (a blend of tracking and control costs) by 8%. We suppose the system parameters can vary by ±30% from their nominal values. Figure 5 compares the states of the closed-loop system with the desired trajectories when the system parameters vary. The MATLAB code used to generate these results is available at http://embeddedlab.csuohio.edu/prosthetics/research/robust-adaptive.html [2].

Hybrid Function Approximation-Based Impedance Control for Prosthetic Legs. In our previous research [7] we developed a process to combine the different control schemes of a prosthesis test robot and a prosthetic leg to yield a stable system. We assumed that the prosthesis test robot was controlled with Slotine and Li's regressor-based controller while the prosthesis was controlled with a regressor-free controller. We addressed this problem by defining a framework within which two controllers could be systematically combined by maintaining their indirect dependence on each other, and we developed a theorem that proved that the combined robotic system was stable, and we showed the efficacy of the system using simulation results. However, the goal of the controllers in [7] was pure motion tracking in the presence of external disturbances. As a result, we obtained good reference trajectory tracking but relatively high control signal magnitudes.

In an effort to reduce the effects of external disturbances, or ground reaction forces (GRFs), we augment impedance control, which is a form of environmental interaction

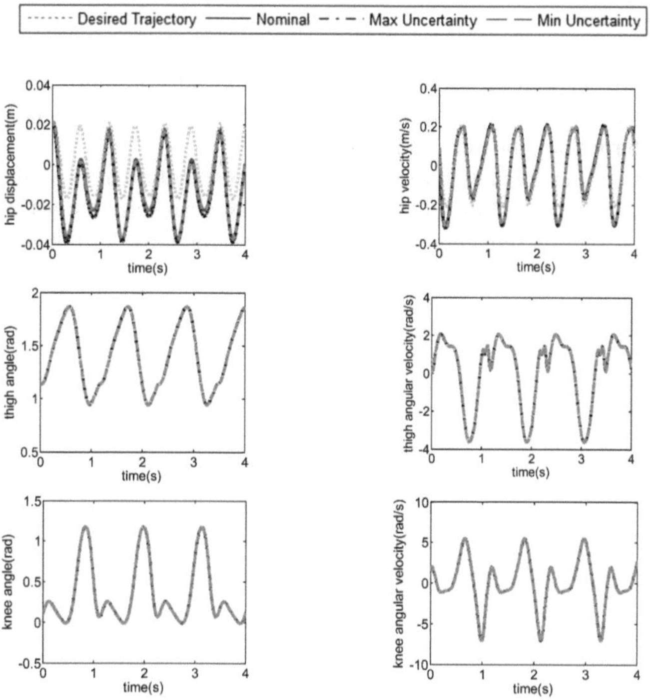

Fig. 5. Joint displacements and velocities with stable robust adaptive impedance control

control, to our previously developed hybrid control scheme. Impedance control gives the modified control scheme the ability to trade off trajectory tracking with control signal magnitude, depending on the nature of the GRFs.

The combination of the prosthesis test robot and the prosthetic leg can be described by the dynamic equations of a rigid robot under the influence of external forces [29, 39]. We use pure motion tracking as the goal of the prosthesis test robot and impedance control as the goal of the prosthetic leg. For impedance control we design a controller such that the closed-loop system behaves like the target impedance

$$M_i(\ddot{q}_r - \ddot{q}_d) + B_i(\dot{q}_r - \dot{q}_d) + I_i(q_r - q_d) = -T_e \tag{7}$$

where $q_r \in \mathbb{R}^n$ and $q_d \in \mathbb{R}^n$ are the reference and desired trajectory respectively, $M_i \in \mathbb{R}^{n\times n}$, $B_i \in \mathbb{R}^{n\times n}$, and $I_i \in \mathbb{R}^{n\times n}$ are the apparent inertia, damping, and stiffness respectively, and $T_e \in \mathbb{R}^n$ captures the external torques and forces applied to the coupled robotic system.

We use MATLAB/Simulink to simulate the system's behavior with the proposed controller, which is a combination of regressor-free environmental interaction control and regressor-based pure motion tracking control; see Figs. 6 and 7. Figure 6 shows good tracking of the reference trajectories for hip displacement, hip angle, and knee angle. The controller for the knee angle gives reasonable trade-offs in tracking and control signal magnitude when under the influence of GRFs, and then resumes tight reference trajectory tracking when there is no GRF.

In Fig. 7 we see that the control signal magnitudes for the hip displacement, hip angle, and knee angle are relatively low, and the vertical GRF is comparable to that experienced during able-bodied walking.

In conclusion, the simulation results show the combination of two different robotic systems with different control schemes. The simulation results show that impedance control is valuable in reducing control signal magnitudes, and hence preventing damage to equipment and reducing strain on amputees.

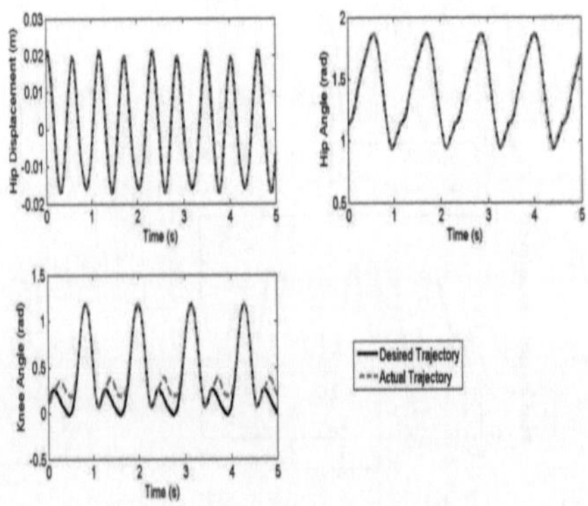

Fig. 6. Joint angle trajectories with hybrid function approximation-based impedance control

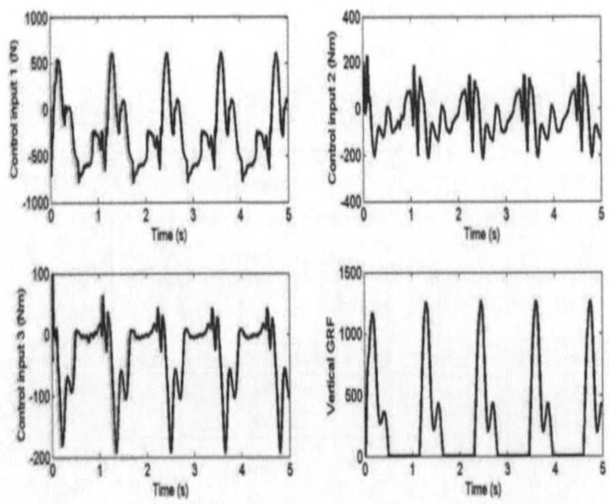

Fig. 7. Control signals and vertical ground reaction force with hybrid function approximation-based impedance control

System Identification and Control Optimization of a Prosthetic Knee. In this research, an EMG-30 geared DC motor is installed in a Mauch SNS knee to create an active prosthesis. The Mauch SNS knee is a widely used passive prosthesis; we adapted it here by detaching the damper connection and driving it with a DC motor.

Our work provides a basic framework for system identification, control optimization, and implementation of an active prosthetic knee during swing phase. To apply velocity control to the system, proportional-integral-derivative control (PID) is utilized due its applicability to an extensive variety of systems, its simplicity, and its ease of use with embedded systems technology. The objective of this research is parameter investigation for PID with respect to prosthetic leg shank length. To accomplish this objective, we first need to develop a prosthetic leg model. We use heuristic algorithms and gradient descent algorithms to identify model parameters and tune the PID controller.

Particle swarm optimization, biogeography-based optimization (BBO) and sequential quadratic optimization (SQP) [18, 20, 32, 37] are selected for identification and tuning. The reason for using more than one optimization algorithm is to avoid local minimum solutions, to discover which algorithm is superior for this task, and to discover how sensitive each heuristic algorithm is to its own parameters.

Hardware setup includes a servo system composed of a desktop PC connected to a Quanser© DAQ card, Matlab with Quanser Quarc software for real-time connectivity, and DAQ hardware; see Fig. 8. The DAQ system delivers an analog control signal to a servo amplifier to drive the EMG30 DC motor. An axial quadrature encoder sends signals through two digital channels. A Mauch SNS knee is attached to an EMG-30 geared DC motor to comprise our active leg prosthesis.

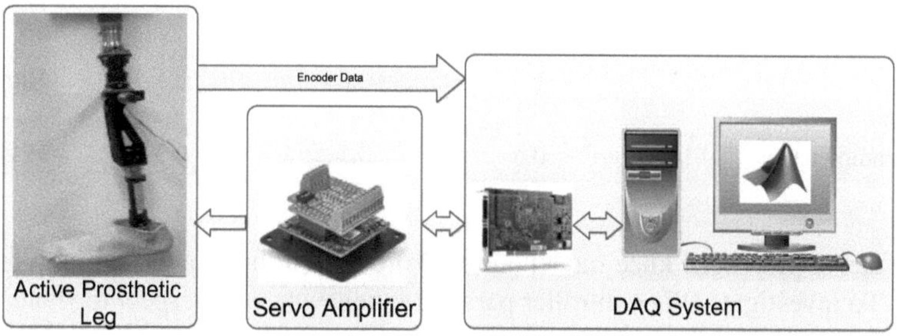

Fig. 8. Hardware setup for system identification and control optimization

Numerical differentiation is usually used to obtain angular velocity by differentiating the encoder signal [42]. This technique often leads to a highly distorted signal due to encoder resolution. Therefore, a Kalman filter is designed to estimate the angular velocity. The DC geared motor and the Mauch SNS are described mathematically in [10]. Matlab Simulink is used to build simulation models. In order to identify model parameters and obtain average performance metrics, optimization algorithms execute 20 times each. The DC motor model and Mauch knee joint model are combined to form the active prosthetic leg model.

In order to see how sensitive BBO and PSO performances are, a sensitivity analysis test is carried out for each algorithm. We say that an algorithm's sensitivity to one of its parameters is "High," "Medium," or "Low," if a given percentage parameter variation leads to a deviation from the best solution by less than 10%, 10–25%, or more than 25%, respectively. Tables 2 and 3 show the algorithm parameter values and their sensitivities.

Table 2. BBO algorithm parameter sensitivity

	Lowest value	Highest value	Test increment	Best value	Sensitivity
Number of generations	50	125	25	**100**	Medium
Population size	50	100	10	**60**	Medium
Mutation probability	5%	20%	5%	**10%**	High
Number of elites	1	5	1	**2**	High

Table 3. PSO algorithm parameter sensitivity

	Lowest value	Highest value	Test increment	Best value	Sensitivity
Number of generations	50	250	50	**100**	Medium
Population size	50	100	10	**60**	Low
Correction factor	0.5	3	0.5	**2**	Medium
Acceleration constant	1	5	1	**0.5**	High
Cognitive parameter	0.05	1	0.05	**0.1**	High
Social parameter	0.1	0.5	0.1	**0.3**	High

The active prosthetic knee model and PID are used to build a closed-loop feedback system. To investigate PID controller parameter behavior with respect to shank length, we use the optimization algorithms to tune the controller parameters (K_p, K_i and K_d). Results show that for model parameter identification, particle swarm optimization gives the best optimization results, and BBO gives better average overall performance than SQP. For PID tuning, BBO achieves the best average overall performance, but PSO shows the fastest average convergence. Finally, we see that increasing shank length results in an increase in the optimal proportional gain, and a decrease in the optimal differential and integral gains; see Fig. 9.

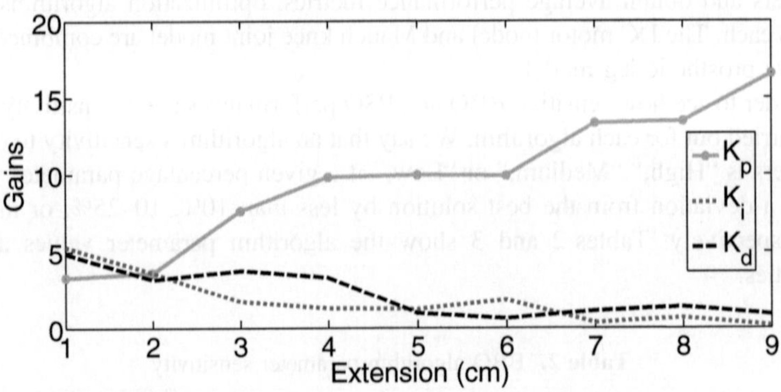

Fig. 9. PID parameter values with respect to shank length

Ground Reaction Force Estimation with an Extended Kalman Filter. A method to estimate GRF in a robot/prosthesis system is presented. The system includes a robot that emulates human hip and thigh motion, and a powered prosthesis for transfemoral amputees, and includes four degrees of freedom: vertical hip displacement, thigh angle, knee angle, and ankle angle. A continuous-time extended Kalman filter (EKF) [38] estimates the states of the system and the GRFs that act on the prosthetic foot.

The system includes eight states: q_1 is vertical hip displacement, q_2 is thigh angle, q_3 is knee angle, q_4 is ankle angle, and their derivatives. Horizontal and vertical GRF is applied to the toe and heel of a triangular foot. The ground stiffness is modeled to simulate GRF. The initial state $x(0)$ is obtained from reference data, and we randomly initialize the estimated state $\widehat{x}(0)$ to include estimation error. The diagonal covariance matrices of the continuous-time process noise and measurement noise are tuned to obtain good performance. Results are shown in Fig. 10. Even with significant initial estimation errors, the EKF converges to the true states quickly.

The performance of the EKF may deteriorate significantly with modeling uncertainties. The H_∞ filter [38] was designed to improve the robustness of state estimation in the presence of modeling errors. Here we assume that the robot/prosthesis system parameters vary by $\pm 10\%$ from their nominal values. In this test the initial value of the state vector $x(0)$ and the measurement and process noise covariances R and Q are identical to those used in the EKF. We also set the initial value of the estimated state vector $\widehat{x}(0)$ to provide an arbitrary but nonzero initial estimation error. The tuning parameters in the H_∞ filter are chosen by the trial and error to obtain good performance. The RMS estimation errors of the two filters are compared in Tables 4 and 5 when $\pm 10\%$ uncertainty on the system parameters is imposed. We see that the H_∞ filter generally performs better than the EKF. However, the EKF still works well.

Table 4. Comparison between EKF and H_∞ filters in terms of RMSE when the system parameters (SPs) deviate from their nominal values by +10%

SPs	$x_1, (m)$	$x_2, (rad)$	$x_3, (rad)$	$x_4, (rad)$	$x_5, (m/s)$	$x_6, (rad/s)$
EKF	0.005	0.007	0.03	0.02	0.06	0.12
H_∞	0.001	0.005	0.01	0.01	0.03	0.12
SPs	$x_7, (rad/s)$	$x_8, (rad/s)$	$x_9, (N)$	$x_{10}, (N)$	$x_{11}, (N)$	$x_{12}, (N)$
EKF	0.47	1.09	4.1	19	7.8	33
H_∞	0.44	0.36	1.5	7.5	5.7	25

Table 5. Comparison between EKF and H_∞ filters in terms of RMSE when the system parameters (SPs) deviate from their nominal values by −10%

SPs	$x_1, (m)$	$x_2, (rad)$	$x_3, (rad)$	$x_4, (rad)$	$x_5, (m/s)$	$x_6, (rad/s)$
EKF	0.007	0.005	0.03	0.02	0.07	0.14
H_∞	0.002	0.005	0.01	0.03	0.03	0.11
SPs	$x_7, (rad/s)$	$x_8, (rad/s)$	$x_9, (N)$	$x_{10}, (N)$	$x_{11}, (N)$	$x_{12}, (N)$
EKF	0.45	1.10	5.5	19	6.2	31
H_∞	0.44	0.44	2.8	9.3	4.5	22

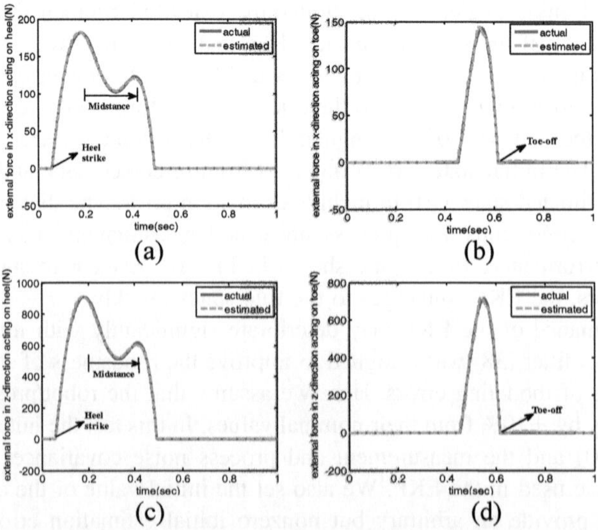

Fig. 10. Horizontal and vertical ground force (GRF) estimation

Electronic Energy Converter Design for Regenerative Prosthetics. Prosthetic models use ideal electromechanical actuators for knee joints, which do not include energy regeneration. In order to focus on energy regeneration, a voltage source converter is designed in this research to interface an electric motor to a supercapacitor.

The converter is designed to resemble a typical H-bridge motor driver. The voltage converter control system allows power to flow from the motor to the capacitor (motor mode) and from the capacitor to the motor (generator mode). During motor mode, the voltage converter's control system modulates the voltage applied to the motor using two circuits; one with the capacitor connected (powering the motor from the capacitor) and one with the capacitor disconnected (shorting the motor connection through the H-bridge). During generator mode, the voltage converter control system changes the impedance connected to the motor using two circuits; one with the capacitor connected (charging the capacitor) and one with the capacitor disconnected (allowing the motor to move with less resistance from the electronics). The circuit and motor are modeled with state space equations using MATLAB and Simulink software.

The converter is augmented to a previously developed mechanical prosthesis model [21]. The model includes the mechanical dynamics of the prosthesis, a ground contact model, and a robust tracking/impedance controller. The controller calculates desired joint torques to achieve trajectory tracking of abled-bodied reference data. The converter replaces the ideal knee motor actuator in this prosthesis model. Since torque and current are proportionally related through the motor dynamics, the tracking/impedance controller is modified to command a desired current that the converter applies through the knee motor to create the desired torque at the knee joint.

A neural network creates an inner control loop for the converter to generate the commanded motor current while the tracking/impedance controller determines torques to meet the tracking and impedance goals for the prosthesis. The neural network includes an input node which compares the motor current generated by the converter to the desired motor current commanded by the tracking/impedance controller. In addition to the error signal of the motor current, a measured ground reaction force input node is used to determine if the prosthesis is in a stance phase or swing phase, a measured velocity input node is used to determine the direction of motor rotation, and a measured torque input node is used to indicate whether the prosthesis is operating in the motoring or generating mode. The neural network contains a single output node which commands a change in duty cycle for the converter to modulate the power flow between the capacitor and the knee motor to achieve the motor torque commanded by the tracking/impedance controller.

The controller gains for the tracking/impedance controller and neural network controller, as well as the physical parameters such as the capacitance of the capacitor and the length of the prosthesis transmission links, are optimized with BBO. The system is optimized using a single set of desired trajectories (hip displacement, hip rotation, knee rotation, and ankle rotation). The control gains and physical parameters selected by BBO achieve the optimization objective of knee angle tracking with a root mean square (RMS) error of 0.13° during five seconds of simulated walking. The selected control gains and physical parameters are then used with test sets of walking data to ensure that the system can maintain tracking with different trajectories. The prosthesis maintains knee angle tracking with an RMS error of 1° with seven different sets of data during five seconds of walking. One representative set of data is shown in Fig. 11. Some data sets show a loss of energy in the capacitor, but the best case results in an increase of 67 Joules of energy during five seconds of walking. The transfer of energy between the knee and capacitor is shown in Fig. 12. It is observed that the capacitor charges as the knee motor produces excess energy and the capacitor discharges as the knee motor consumes energy.

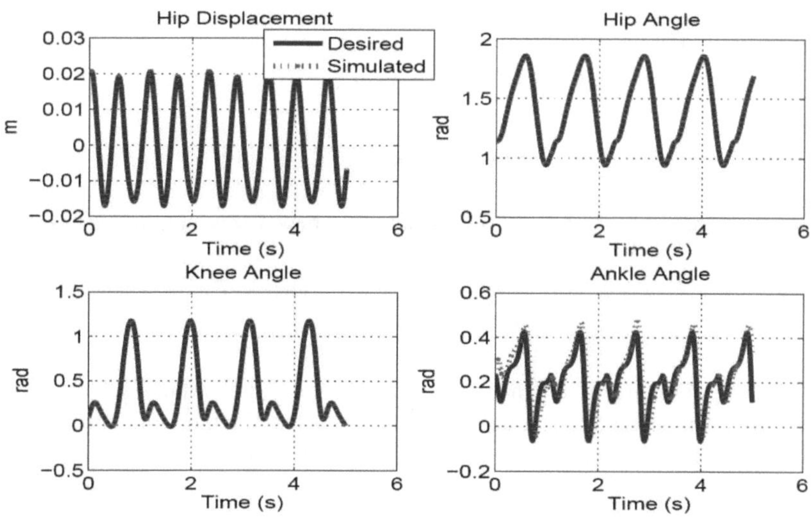

Fig. 11. Reference and simulated trajectories of hip displacement, hip rotation, knee rotation, and ankle rotation during five seconds of walking

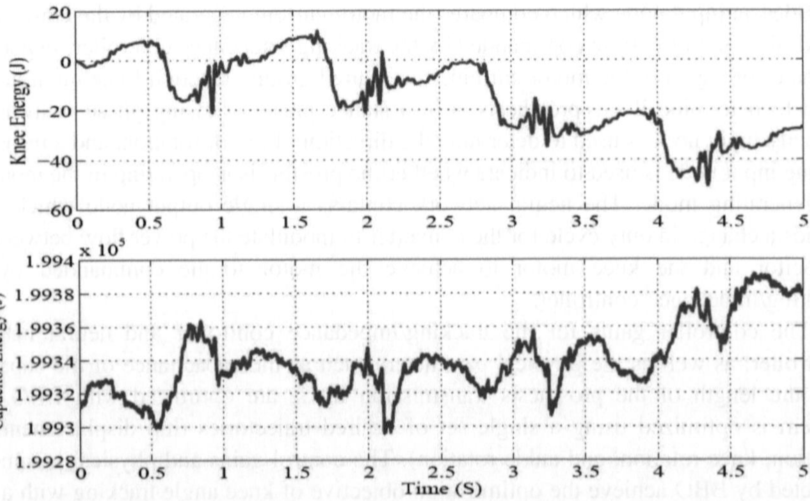

Fig. 12. Comparison between energy produced at the knee and energy stored in the capacitor

Fuzzy Logic for Robot Navigation. This research uses fuzzy logic to find a path for a mobile robot to navigate in an environment with both static and dynamic obstacles when the robot does not have any prior information about the obstacle locations. The robot stores the coordinates of previously visited locations in memory to avoid getting stuck in dead ends.

The robot radar returns a fuzzy set based on the distance L_i from obstacle i (see Fig. 13): $\mu_i^{\varphi}(\varphi_i) = \frac{L_i}{L_{max}}$. The robot finds the angle between its position and the target position, which we call α. If the robot moved in the α direction in an obstacle-free environment it would follow a direct line to the target. However, there are obstacles in the path. To find a safe path around the obstacles, we introduce a Gaussian fuzzy set [13, 14, 28, 43] which has a maximum value at α:

$$\mu_i^{\alpha}(\varphi_i) = e^{-\left(\frac{(\varphi_i-\alpha)^2}{2\sigma^2}\right)}.$$

We combine $\mu_i^{\varphi}(\varphi_i)$ and $\mu_i^{\alpha}(\varphi_i)$ to obtain a new fuzzy set, $\mu_i^{\psi}(\psi_i)$, shown in Fig. 13.

$$\mu_i^{\psi}(\varphi_i) = \min(\mu_i^{\alpha}(\varphi_i), \mu_i^{\varphi}(\varphi_i))$$

The movement direction is the maximum point in $\mu_i^{\psi}(\psi_i)$, which we call φ_A. If the robot moves with angle of φ_A, it will touch the obstacles. We therefore introduce a new fuzzy set that has the value 1 in a range of 120° around φ_A:

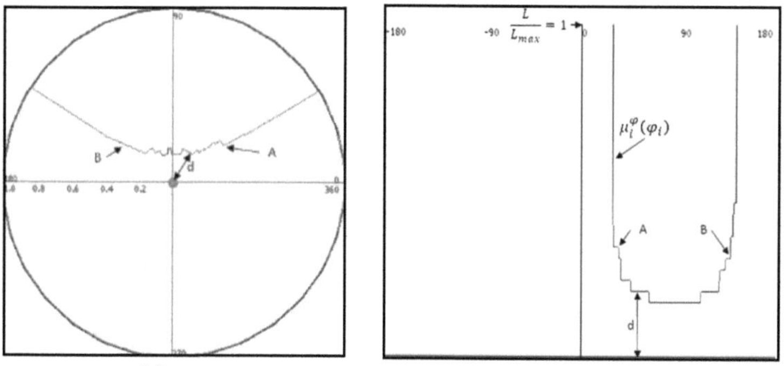

(a) (b)

Fig. 13. (a) A polar radar map in the presence of an obstacle, and (b) its transformation to Cartesian coordinates

$$\mu_1^{\theta}(\varphi_i) = \begin{cases} 1 & |\varphi_i - \varphi_A| < 60 \\ 0 & \text{otherwise} \end{cases}$$

In the next step we defuzzify $\mu^{\psi}(\varphi_i) * \mu_1^{\theta}(\varphi_i)$ using center of mass defuzzification [31], which is shown in Fig. 14.

Simulations confirm that the proposed approach provides reliable navigation. However, the robot is only able to get to the target point if it does not enter a dead end zone. Examples of dead end zones include rooms, single-entry areas, and other situations where the robot needs to move backward to find the path to the target. In practical applications, we cannot guarantee that a map won't have dead end zones.

The robot therefore saves visited paths in memory. The data in the robot's memory is a list of coordinates which we call Memory Points (*MPs*). The robot should avoid visiting the same place multiple times and it should be able distinguish between locations that were visited in the recent and in the distant past. To achieve this goal, we assign weights to each coordinate in memory (w). Weights change based on the distance to the robot. Their values are calculated with a derivative of a Gaussian distribution as shown in Fig. 15 and as described by the following:

$$w_i(MP_i) = \left| \frac{-de^{-\left(\frac{d^2}{2\sigma^2}\right)}}{\sigma} \right|, \quad \text{where} \begin{cases} d = \sqrt{\left(x_{MP_i} - x_{RP_i}\right)^2 + \left(y_{MP_i} - y_{RP_i}\right)^2}, \\ i = 1, 2, \ldots, N \end{cases}$$

N is the number of *MP*s in memory and *RP* is the current position of the robot. With this weighting function, the robot is more likely to visit very recent and very old locations and tends to avoid coordinates that are in between.

In different layouts and different robot and target positions, the robot can find a path to the target point successfully; see Fig. 16.

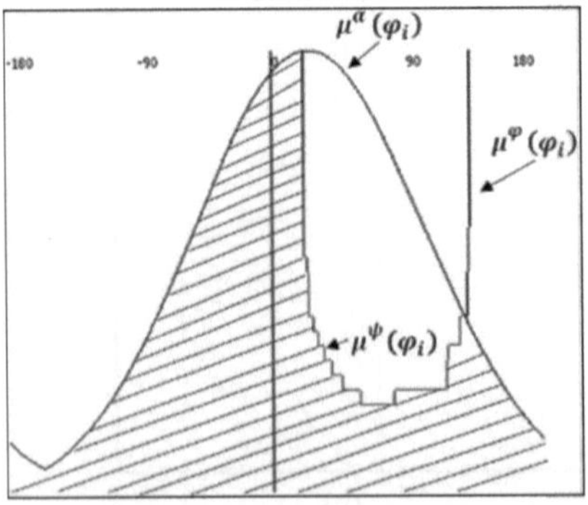

Fig. 14. Shaded area is $\mu_i^{\psi}(\psi_i)$

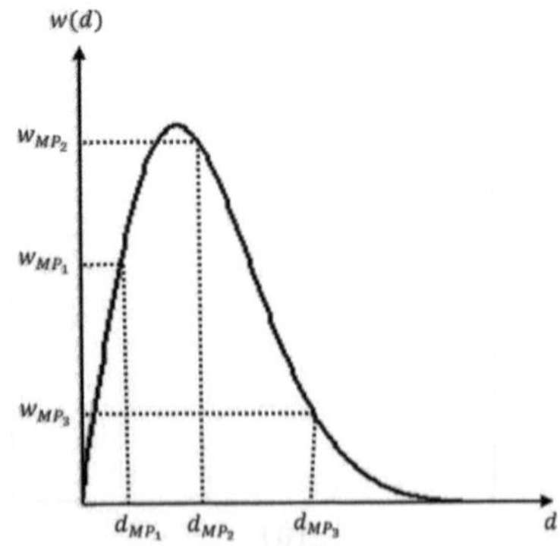

Fig. 15. Weight distribution of Memory Points for robot path planning

5 Education and Research Integration

The seven research projects described in the preceding section have a common educational core which enables their success. This section reviews those common elements.

First, most of the students involved in research come to CSU with the promise of multi-year research funding. Most funding comes from external agencies, such as the US government or industry. Some funding comes from the limited resources within CSU. In order to be productive researchers, students need to be devoted full-time to their studies and research, and they need to be worry-free with regards to finances. Successful research depends on funding, and funding depends on faculty.

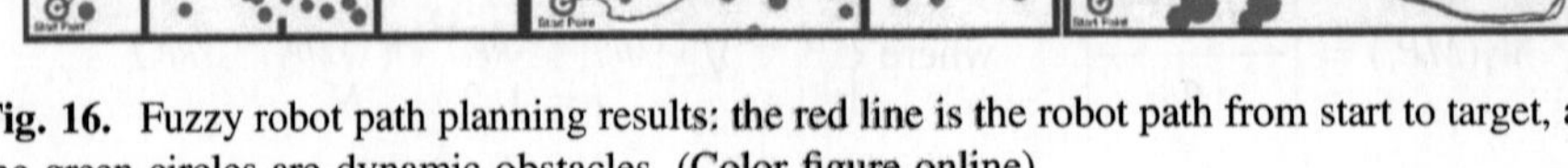

Fig. 16. Fuzzy robot path planning results: the red line is the robot path from start to target, and the green circles are dynamic obstacles. (Color figure online)

Second, research students are involved not only in coursework and research, but also in teaching-assistant duties. Some students assist with labs, tutoring, and grading, while other students are given sole responsibility for an entire undergraduate course, depending on their experience level. Superficially, this detracts from their research. However, in the long run, it enhances their research by providing them with training in the area of communication skills, problem solving, and networking.

Third, most research students are recruited as doctoral students, who are more serious about research than master's students. Some students enter CSU as master's students, but always with the intent of ultimately progressing to doctoral studies.

Fourth, research students are required to attend and present at weekly research seminars at CSU. These seminars provide the students with opportunities to learn from each other and from faculty, to network with each other, and to practice their communication skills. This indicates the need for a critical mass of research students in order to ensure a successful research enterprise.

Fifth, research students are required to publish and present their research at one or more international conferences each year. This provides them with similar benefits as the weekly seminars discussed above, but at a larger scale.

Sixth, research students with varying levels of experience and disciplinary focus all work together. Research participants include faculty, post-doctoral scholars, visiting scholars, doctoral students, master's students, and undergraduate students. Diversity also intentionally includes gender, ethnicity, and nationality. This diversity allows research students to be involved in both receiving and providing mentoring, and in learning to work across comfortable boundaries.

Seventh, research students are given as much responsibility as possible in the conduct of their research, and this responsibility gradually increases as the students gain experience. Faculty advisors fill the role of advising, but generally try to keep a hands-off approach in the daily conduct of student research. This approach teaches students to be proactive in solving problems, to take the initiative in seeking advice, and to take responsibility for their research. Faculty advisors are responsible to help students find the right balance so that the students don't go down the wrong research path or stall in their research efforts.

Many educational factors are involved in successful research. The above factors are just a few. Many more could be incorporated at CSU and other universities. But the most important consideration here is that in order to be successful, faculty must take an intentional approach to integrating education and research, and to graduating research students who are prepared to take the lead in the next generation.

6 Conclusions

The authors have described university student training. The description has focused on student participation in the US NSF project "Optimal prosthesis design with energy regeneration" and the application of ICT and modelling technologies.

Several factors play an important role in the results of this chapter. Student research requires skill in programming and software, and a broad theoretical knowledge in computer science, and mechanical, electrical, and control engineering. Students used MATLAB, Simulink, and toolboxes (Optimization, Fuzzy Logic, etc.), and programming in C and C++. The software used for robot trajectory planning research was designed and written by students in C++, and the GUI was designed using Qt and OpenGL. Standard libraries were used to make the software cross-platform.

The most important foundation for student research is theoretical knowledge in fundamental and elective disciplines such as Circuits, Linear Systems, Control

Systems, Nonlinear Control, Machine Learning, Artificial Intelligence, Intelligent Controls, Optimal State Estimation, Optimal Control, Embedded Systems, Robot Modeling and Control, Probability and Stochastic Processes, Population-Based Optimization, and Prosthesis Design and Control, which provides a basic understanding of human biomechanics and lower-limb prosthesis design and control. These courses played a vital role in the proper grounding of basic and advanced ICT and control theory for robotic and prosthetic leg research. The facilities at CSU and funding from the NSF significantly helped in furthering student research-based education.

Finally, student participation in government-sponsored research, student exchanges of research experiences with each other, and publication of research results in high-caliber journals and conferences [1, 2, 7, 11, 18, 29], provide students with effective training and self-confidence in their higher education. Research-based education also allows students to obtain practical experience as research assistants, with corresponding responsibilities in the development and implementation of research projects.

Student participation in real-world research significantly influences their engineering and research qualifications by: (a) giving them a strong understanding of ICT and engineering concepts that are covered in corresponding courses; (b) giving them practical experience and the ability to apply theoretical knowledge; (c) giving them the opportunity to learn technical material independently; (d) helping them improve fundamental skills to apply in other research in their future; (e) providing them with a rich interdisciplinary research environment; and (f) providing them with an understanding of concepts both familiar and unfamiliar. Through extensive literature review and actively seeking ways to solve research problems, students are prepared to make meaningful future contributions to the field of ICT and control engineering.

Acknowledgements. The authors thank the Fulbright Program (USA) for supporting Prof. Y.P. Kondratenko with a Fulbright 2015-2016 scholarship and for making it possible for this team to conduct research in together in the USA. This research was partially supported by US NSF Grant 1344954.

References

1. Azimi, V., Simon, D., Richter, H., Fakoorian, S.A.: Robust composite adaptive transfemoral prosthesis control with non-scalar boundary layer trajectories. In: Proceedings of the American Control Conference (ACC), Boston, MA, USA (2016)
2. Azimi, V., Simon, D., Richter, H.: Stable robust adaptive impedance control of a prosthetic leg. In: Dynamic Systems and Control Conference, Columbus, Ohio, USA, October 2015
3. Bredeweg, B., Forbus, K.D.: Qualitative modeling in education. AI Mag. **24**(4), 35–46 (2003). American Association for Artificial Intelligence
4. Chou, C., Tsai, C.-C.: Developing web-based curricula: issues and challenges. J. Curriculum Stud. **34**(6), 623–636 (2002)
5. Clements, D.H.: Curriculum research: toward a framework for "Research-Based Curricula". J. Res. Math. Educ. **38**(1), 35–70 (2007)
6. Conditt, M.A., Noble, P.C., Thompson, M.T., Ismaily, S.K., Moy, G.J., Mathis, K.B.: A computerized bioskills system for surgical skills training in total knee replacement. Proc. Inst. Mech. Eng. Part H J. Eng. Med. **221**(1), 61–69 (2007)

7. Ebeigbe, D., Simon, D., Richter, H.: Hybrid function approximation based control with application to prosthetic legs. In: IEEE International Systems Conference, Orlando, FL, USA, April 2016
8. Gáti, J., Kártyás, G.: Computer system support for higher education programs in engineering. In: Proceedings of 4th International Conference on Emerging Trends in Engineering and Technology (ICETET), pp. 61–65 (2011)
9. Gilbert, J.K.: Models and modelling: routes to more authentic science education. Int. J. Sci. Math. Educ. **2**(2), 115–130 (2004)
10. Gonçalves, J., Lima, J., Costa, P., Moreira, A.P.: Modeling and simulation of the EMG30 geared motor with encoder resorting to SimTwo: the official Robot@Factory simulator. In: Azevedo, A. (ed.) Advances in Sustainable and Competitive Manufacturing Systems, pp. 307–314. Springer, Heidelberg (2013)
11. Khademi, G., Mohammadi, H., Simon, D., Hardin, E.C.: Evolutionary optimization of user intent recognition for transfemoral amputees. In: IEEE Biomedical Circuits and Systems Conference (BioCAS), Atlanta, Georgia, USA, pp. 1–4 (2015)
12. Kondratenko, Y.P., Kondratenko, G.V.: Robotic system with myoelectric adapting for rehabilitation tasks. ACTA Bioeng. Biomech. **3**, 1–6 (2001)
13. Kondratenko, Y., Kondratenko, V.: Soft computing algorithm for arithmetic multiplication of fuzzy sets based on universal analytic models. In: Ermolayev, V., Mayr, H., Nikitchenko, M., Spivakovsky, A., Zholtkevych, G. (eds.) Information and Communication Technologies in Education, Research, and Industrial Application. Communications in Computer and Information Science, vol. 469, pp. 49–77. Springer, Cham (2014)
14. Kondratenko, Y.P., Klymenko, L.P., Al Zu'bi, E.Y.M.: Structural optimization of fuzzy systems' rules base and aggregation models. Kybernetes **42**(5), 831–843 (2013)
15. Kondratenko, Y., Khademi, G., Azimi, V., Ebeigbe, D., Abdelhady, M., Fakoorian, S.A., Barto, T., Roshanineshat, A.Y., Atamanyuk, I., Simon D.: Information, communication, and modeling technologies in prosthetic leg and robotics research at Cleveland State University. In: Ermolayev, V., et al. (eds.) Proceedings of 12th International Conference on Information and Communication Technologies in Education, Research, and Industrial Application, 21–24 June 2016, Kyiv, Ukraine, ICTERI 2016, vol. 1614, pp. 168–183. CEUR-WS (2016)
16. Kondratenko, Y., Simon, D., Atamanyuk, I.: University curricula modification based on advancements in information and communication technologies. In: Ermolayev, V., et al. (eds.) Proceedings of 12th International Conference on Information and Communication Technologies in Education, Research, and Industrial Application, 21–24 June 2016, Kyiv, Ukraine, ICTERI 2016, vol. 1614, pp. 184–199. CEUR-WS (2016)
17. Louca, L.T., Zacharia, Z.C.: The use of computer-based programming environments as computer modelling tools in early science education: the cases of textual and graphical program languages. Int. J. Sci. Educ. **30**(3), 287–323 (2008)
18. Ma, H., Simon, D.: Biogeography-based optimization with blended migration for constrained optimization problems. In: Proceedings of the 12th Annual Conference on Genetic and Evolutionary Computation, New York, NY, USA, pp. 417–418 (2010)
19. McKee, N.H., Agur, A.M., Tsang, W., Singh, K.S.: Possible uses of computer modeling of the functioning human hand. Clin. Plast. Surg. **32**(4), 635–641 (2005)
20. Mo, H., Xu, Z.: Research of biogeography-based multi-objective evolutionary algorithm. In: Khosrow-Pour, M. (ed.) Interdisciplinary Advances in Information Technology Research, pp. 125–135 (2013)
21. Mohammadi, H., Richter, H.: Robust tracking/impedance control: application to prosthetics. In: American Control Conference, Chicago, Illinois, July 2015 (2015)

22. Musa, A., Al-Dmour, A., Fraij, F., Gonzalez, V., Al-Hashemi, R.: Developing a virtual laboratory for a communication and computer networking course. Int. J. Continuing Eng. Educ. Life Long Learn. **20**(3–5), 390–406 (2010)
23. Nehrir, M.H., Fatehi, F., Gerez, V.: Computer modeling for enhancing instruction of electric machinery. IEEE Trans. Educ. **38**(2), 166–170 (1995)
24. Nemanjic, B., Svetozar, N.: Computer Simulations: Technology, Industrial Applications and Effects on Learning. Nova Science Publishers Inc., New York (2012)
25. Neptune, R.R.: Computer modeling and simulation of human movement. Sci. Principles Sports Rehabil. **11**(2), 417–434 (2000)
26. Panselinas, G.E., Komis, V.: Using educational software to support collective thinking and test hypotheses in the computer science curriculum. Educ. Inform. Technol. **16**(2), 159–182 (2011)
27. Patil, L.S., Patil, K.D., Thosar, A.G.: The role of computer modeling and simulation in power electronics education. In: Proceedings of the 2nd International Conference on Emerging Trends in Engineering and Technology (ICETET), pp. 416–419 (2009)
28. Piegat, A.: Fuzzy Modeling and Control. Springer, Heidelberg (2001). doi:10.1007/978-3-7908-1824-6
29. Richter, H., Simon, D., Smith, W., Samorezov, S.: Dynamic modeling, parameter estimation and control of a leg prosthesis test robot. Appl. Math. Model. **39**, 559–573 (2015)
30. Richter, H., Simon, D.: Robust tracking control of the prosthesis test robot. J. Dyn. Syst. Measur. Control **136**(3), Paper No. DS-13–1052 (2014)
31. Runkler, T.: Selection of appropriate defuzzification methods using application specific properties. IEEE Trans. Fuzzy Syst. **5**(1), 72–79 (1997)
32. Sayed, M., Saad, M., Emara, H., El-Zahab, E.A.: A novel method for PID tuning using a modified biogeography-based optimization algorithm. In: Proceedings of the 24th Chinese Control and Decision Conference (CCDC), pp. 1642–1647 (2012)
33. Schonning, A.: Biomechanical applications of computers in engineering education. In: ASME 2007 International Design Engineering Technical Conferences and Computers and Information in Engineering Conference, Las Vegas, Nevada, USA, pp. 389–393. American Society of Mechanical Engineers (2007)
34. Schwarz, C.V., Meyer, J., Sharma, A.: Technology, pedagogy, and epistemology: Opportunities and challenges of using computer modeling and simulation tools in elementary science methods. J. Sci. Teacher Educ. **18**(2), 243–269 (2007)
35. Shaalan, H.: Using software to enhance power engineering education in a technology program. IEEE Power Eng. Rev. **21**(6), 62–63 (2001)
36. Shih, N.-J.: RP-aided computer modeling for architectural education. Comput. Graph. **30**(1), 137–144 (2006)
37. Simon, D.: Evolutionary Optimization Algorithms: Biologically Inspired and Population-Based Approaches to Computer Intelligence. Wiley, Hoboken (2013)
38. Simon, D.: Optimal State Estimation: Kalman, H-infinity, and Nonlinear Approaches. Wiley, Hoboken (2006)
39. Slotine, J.-J.E., Coetsee, J.A.: Adaptive sliding controller synthesis for non-linear systems. Int. J. Control **43**(6), 1631–1651 (1986)
40. Teodoro, V.D., Neves, R.G.: Mathematical modelling in science and mathematics education. Comput. Phys. Commun. **182**(1), 8–10 (2011)
41. Van Amerongen, J., Breedveld, P.: Modelling of physical systems for the design and control of mechatronic systems. Ann. Rev. Control **27**(1), 87–117 (2003)
42. Yang, W.: Applied Numerical Methods Using MATLAB. Wiley-Interscience, Hoboken (2005)
43. Zadeh, L.A.: Fuzzy sets. Inf. Control **8**, 338–353 (1965)

44. Zhang, B.H., Liu, X.F., Krajcik, J.S.: Expert models and modeling processes associated with a computer-modeling tool. Sci. Educ. **90**(4), 579–604 (2006)
45. Zhang, Q.: GUI based computer modeling and design platform to promote interactive learning in fiber optic communications. In: Proceedings of the 37th Annual Frontiers in Education Conference - Global Engineering: Knowledge Without Borders, Opportunities Without Passports, FIE 2007, pp. S3H-14–S3H-19 (2007)

The Cloud-Based Learning Component with Maxima System and Its Evaluation

Mariya Shyshkina(✉)

Institute of Information Technologies and Learning Tools of the National Academy of Educational Sciences of Ukraine,
M. Berlinskoho Str., 9, Kiev, Ukraine
marple@ukr.net

Abstract. Nowadays, innovative technological solutions for providing access to electronic resources and their configuration within a cloud-based university educational environment have given rise to diverse research trends especially for learning Math and Computer Science disciplines. This article outlines the conceptual framework of the study by reviewing existing approaches for electronic resources access organization within cloud-based settings. The cloud-based learning component providing Maxima system access with the use of the hybrid service model is described and proved. The problems of learning components quality evaluation in the cloud-based learning environment are outlined. Indicators of cloud-based learning resources quality evaluation are proposed. An empirical estimation of the proposed approach and current developments of its implementation are provided.

Keywords: Learning environment · Cloud computing · Electronic resources · Quality · Hybrid model · University

1 Introduction

The processes of innovative development of a learning space, which is created in educational institutions, are to account for the tendencies of learning technologies improvement on the basis of emerging ICT. Cloud computing (CC) technology is used to enhance multiple access and joint use of educational resources at different levels and domains, combining the corporate resources of the university and other learning resources within a united framework. Progress in the area has provided new insights into the problems of educational electronic resources provision and configuration within the learning environment, bringing new models and approaches. The promising trend of research is concerned to the hybrid service models [2, 10, 15]. It has given rise to research of better ways of introducing innovative technology into the process of learning Math and Computer Science disciplines.

A set of different service models may be elaborated and combined to provide access for the cloud-based learning components of mathematical software. There are numerous kinds of powerful and multifunctional software, so called the Systems of Computer Mathematics (SCM), that are widely used now in the sphere of Math and Computer Science learning. To provide access to SCM in the cloud-based learning

A. Ginige et al. (Eds.): ICTERI 2016, CCIS 783, pp. 156–168, 2017.
https://doi.org/10.1007/978-3-319-69965-3_9

environment is the matter of its technological design and pedagogical estimation. Quality evaluation of the learning components focused on the different SCM application is a promising way to choose and approve the most appropriate delivery settings.

The purpose of the article is analysing the ways of design and application of the cloud-based component providing Maxima system access in the learning process, substantiating and validating the indicators and approach to its evaluation and also to consider the possible ways of its use in a wider context.

The *research method* involved analysing the current research (including the domestic and foreign experience of the application of cloud-based learning services to reveal the state of the art and research indicators), examining existing models and approaches, considering existing methods of learning resources evaluation and its adaptation for use within the cloud-based settings, conducting pedagogical experiments, expert evaluations.

2 Problem Statement

The challenges of making the ICT-based learning components of the university environment fit the needs of its users, have led to the search for the most reasonable ways for its design and delivery within the cloud-based settings. The cloud-based components possess many progressive features including better adaptability and mobility, as well as full-scale interactivity, free network access, a unified structure among others [2, 12, 14]. So, the modeling and analysis of their design and deployment in view of the current tendencies of modern ICT advance and available learning experience have come to the fore.

Among the priority issues there are those concerning existing approaches and models for electronic educational resources delivery within the hybrid cloud-based setting; the cloud-based learning components quality assessment techniques; quality research indicators substantiation and validation; evaluation of current experience of cloud-based models and components use.

3 State of the Art

According to the recent research [2, 5, 8, 11, 12], the problems of implementing cloud technologies in educational institutions so as to provide software access, support collaborative learning, implement scientific and educational activities, support research and project development, exchange experience are especially challenging. The formation of the cloud-based learning environment is recognized as a priority by the international educational community [9], and is now being intensively developed in different areas of education, including mathematics and engineering [1, 4, 18, 20].

The transformation of the modern educational environment of the university by the use of the cloud-based services and cloud computing delivery platforms is an important trend in research. The topics of software virtualization and the forming of a unified ICT infrastructure on the basis of CC have become increasingly popular lines of investigation [4, 11]. The problems with the use of private and public cloud services, their advantages and disadvantages, perspectives on their application, and targets and implementation strategies are within the spectrum of this research [3, 4, 18].

There is a gradual shift towards the outsourcing of ICT services that is likely to provide more flexible, powerful and high-quality educational services and resources [2]. There is a tendency towards the increasing use of the software-as-a-service (SaaS) tool. Along with SaaS the network design and operation, security operations, desktop computing support, datacentre provision and other services are increasingly being outsourced as well. Indeed, the use of the outsourcing mechanism for a non-core activity of any organization, as the recent surveys have observed happening in business, is now being extended into the education sector [5]. So, the study of the best practices in the use of cloud services in an educational environment, the analysis and evaluation of possible ways of development, and service quality estimation in this context have to be considered.

The valuable experience of the Massachusetts institute of technology (MIT) should be noted in concern to the cloud based learning environment formation in particular as for access to mathematical software. The Math software is available in the corporate cloud of the University for the most popular packages such as *Mathematica, Mathlab, Maple, R, Maxima* [21]. This software is delivered in the distributed mode on-line through the corporate access point. This is to save on license pay and also on computing facilities. The mathematics applications require powerful processing so it is advisable to use it in the cloud. On the other case the market need in such tools inspires its supply by the SaaS model. This is evidenced by the emergence of the cloud versions for such products as Sage MathCloud, Maple Net, MATLAB web-server, WebMathematica, Calculation Laboratory and others [1, 4]. Really there is a shift toward the cloud-based models as from the side of educational and scientific community, and also from the side of product suppliers. The learning software actually becomes a service in any case, let it be a public or a corporate cloud.

Recently there has been a trend towards convergence and integration of various mathematical packages. For example, the latest versions of Mathematica and Maple are supplied by powerful tools for visual programming; MathCAD can work together with MatLab etc. So, for the aim of practical training any of the above packages may be used with regard to specific traditions and support opportunities of education institutions. These factors significantly influence the choice of software that can be installed “in the cloud”.

An essential feature of the cloud computing conception is dynamical supply of computing resources, software and hardware its flexible configuration according to user needs. So comparison of different approaches and cloud models of software access is the current subject matter of educational research [3, 4, 17, 19]. Despite of the fact that the sphere of CC is rather emerging there is a need of some comparison of the achieved experience to consider future prospects [19]. Also the problems of software choice in the learning complexes to be implemented in a cloud arise. This leads to the problems of cloud-based learning resources quality evaluation techniques and research indicators substantiation.

In the cloud-based learning environment, new ways of electronic educational resources (EER) quality control arise. There are specific forms of the organization of learning activity related to quality estimation. For example there are e-learning systems based on the modelling and tracking of individual trajectories of each student's progress, knowledge level and further development [22]. This presupposes the adjustment, coordination of training, consideration of pace of training, diagnosis of achieved level of mastery of the material, consideration of a broad range of various facilities for learning to ensure suitability for a larger contingent of users. The vast data collections about the students' rates of learning are situated and processed in the cloud [22]. There are also collaborative forms of learning where the students and teachers take part in the process of resource elaboration and assessment; this is possible in particular by means of the SageMathCloud platform [1].

The special attention is to be paid to the system Maxima, because it is easy to master, it is comparable to such systems as Maple and Mathematica as for solving problems (for example in the field of operations research) and is free accessed. It is equipped with a menu system that enables character conversion, solve equations, compute derivatives, integrals, etc., avoiding some additional efforts as for learning the special language tools to implement these actions. In view of that the Maxima system can be used to study Math and Computer Science disciplines even on the first year of study at the pedagogical university [15]. The use of Maxima will not cause any difficulties for the students as for solving problems of mathematical analysis and linear algebra – the students are required only to choose a menu item and enter the expression. However, programming within Maxima requires knowledge of certain language and syntax, as well as some commands.

Thus, in view of the current tendencies, the research questions are: how can we take maximum advantage of modern network technologies and compose the tools and services of the learning environment to achieve better results? What are the best ways to access electronic resources if the environment is designed mainly and essentially on the basis of CC? What are the most reasonable approaches to validate quality evaluation criteria? This brings the problem of the cloud-based learning components modelling, evaluation and design to the forefront.

4 The Method of Cloud-Based Learning Resources Evaluation

The number of electronic educational resources is increasing and this trend is likely to intensify. By means of CC-based tools, a significant lifting of restrictions on the implementation of access to qualitative leaning resources may be achieved. Now, these questions are not a matter of future perspective, they need practical implementation. For this purpose, the problem of the design and delivery of electronic educational resources in the cloud-based environment in particular within the hybrid service modes is a complex one and not only should technological needs be considered, but also the pedagogical aspects [17].

Due to the significant educational potential and novel approaches to environmental design, its formation and development, these questions remain the matter of theoretical and experimental studies, the refinement of approaches, and the search for models, methods and techniques, as well as possible ways of implementation [2].

To carry out research and experimental activities and the implementation and dissemination of the results, the Joint research laboratory of the Institute of Information Technologies and Learning Tools of NAES of Ukraine and the Kherson State University was created in 2011 with the focus on issues of educational quality management using ICT [23].

The method of electronic resources quality estimation was developed and used in the Joint laboratory of EER quality control [7, 23]. In this case, the different quality parameters have been detailed and selected [7]. It is important that the psychological and pedagogical parameters are estimated in the experimental learning process, while the other types of parameter such as technological or ergonomic may be estimated out of this process.

The prospective way of the estimation of the quality of learning resources is by means of the cloud-based environment. As the resources are collectively accessed, there is a way to allow experts into the learning process so they may observe and research their functioning. This is a way to make the process of quality estimation easier, more flexible and quicker. The process of estimation becomes anticipatory and timely. The estimation may be obtained just once along with the process of EER elaboration, and it is very important to facilitate the process [16].

There are several groups of quality criteria to be taken into consideration and checked in the process of complex quality assessment of ICT-based learning tools. Generally, there are main groups such as: psychological and pedagogical indicators; and ergonomic and technological indicators [7, 23]. There is no single set of criteria clearly acknowledged to be unambiguous. There is a problem of criteria substantiation as the didactic and methodical, and psychological aspects of educational use of ICT, are hardly regulated and standardized [13]. Still, there are research works devoted to the problems of quality evaluation in this field, where the system of quality criteria has been substantiated and proved experimentally [7].

Another kind of problem is connected to the cloud-based learning tools' quality estimation. Specific kinds of criteria are valuable in this case with regard to innovative features of advanced learning settings. So, for the purpose of this study, the quality criteria were scrutinized to reveal the most significant. There were also two groups of indicators selected: the pedagogical and psychological; and technological. It is not feasible to take into account all possible criteria inherent to cloud-based tools' application. There are a lot of technical and technological aspects to be considered, such as portability, sustainability, security, and others. Not to underestimate the importance of all relevant features, the study is concerned with those quality aspects that are valuable only in the case of educational use of ICT-based tools.

Therefore, among the variety of technological parameters, this study focuses on those that are important for pedagogical study in relation to the introduction of emerging ICT. Technological innovations cause shifts in pedagogical approaches and transformations of target, content, and methodological aspects of educational systems [2]. So, the quality of emerging ICT tools must reflect the prospected shift in learning technologies and improvement of pedagogical outcome.

Among the technological indicators of cloud-based educational resource quality evaluation, are those concerned with *ease of access,* showing if electronic resource access organization is easy and convenient; *the intuitive clarity of the interface,* reflecting if the user interface is clear and easy to learn; *responsiveness*, meaning performance in real time work; *sustainability,* concerning capability of functioning while working with the resource from any computer via a browser; *support of collaborative work,* encompassing facilities to support collaboration in the learning process; *ease of integration,* meaning suitability to be incorporated into a single environment along with other resources; and *usefulness,* covering overall utility (feasibility) of resource use.

The pedagogical and psychological criteria of the cloud-based learning resource quality evaluation should be the same as any other educational electronic resource in many respects. However, this is an important step in the quality evaluation process that cannot be neglected while investigating any tool aimed at learning. The set of psychological and pedagogical indicators for the research are as follows: the *scientific clarity* of the content; *accessibility* of the content, delivered by the resource; *fostering the intellectual development* of a learner while working with the resource; *problem orientation* of content and functioning; *personalization* in the learning process; *adaptability* as suitability for most of the possible user contingent; *methodical usefulness,* meaning the most appropriate support for learning methods; *professional orientation* as providing learners' professional development; and *feedback* connection.

As the criteria set is identified and formed, there is a need to provide validity of every indicator that may be achieved by this research.

5 The Design of the Cloud-Based Learning Component with the Use of the Maxima System

To research the hybrid service model of learning software access, especially for the mathematical software a joint investigation was undertaken in 2013–2014 at the Institute of Information Technologies and Learning Tools of the NAES of Ukraine and Drohobych State Pedagogical University named after I.Franko. At the pedagogical experiment the cloud-based learning component with Maxima system was designed and used for Operations Research study.

In this case, the implementation of software access due to the hybrid cloud deployment was organised.

At the pedagogical university, the experimental base was established where the cloud version of the Maxima system (which is mathematical software), installed on a virtual server running Ubuntu 10.04 (Lucid Lynks), was implemented. In the repository of this operational system is a version of Maxima based on the editor Emacs, which was installed on a student's virtual desktop [15].

In Fig. 1, the configuration of the virtual hybrid cloud used in the pedagogical experiment is shown. The model contains a virtual corporate (private) subnet and a public subnet. The public subnet can be accessed by a user through the remote desktop protocol (RDP). In this case, a user (student) refers to certain electronic resources and a computing capacity set on a virtual machine of the cloud server from any device, anywhere and at any time, using the Internet connection.

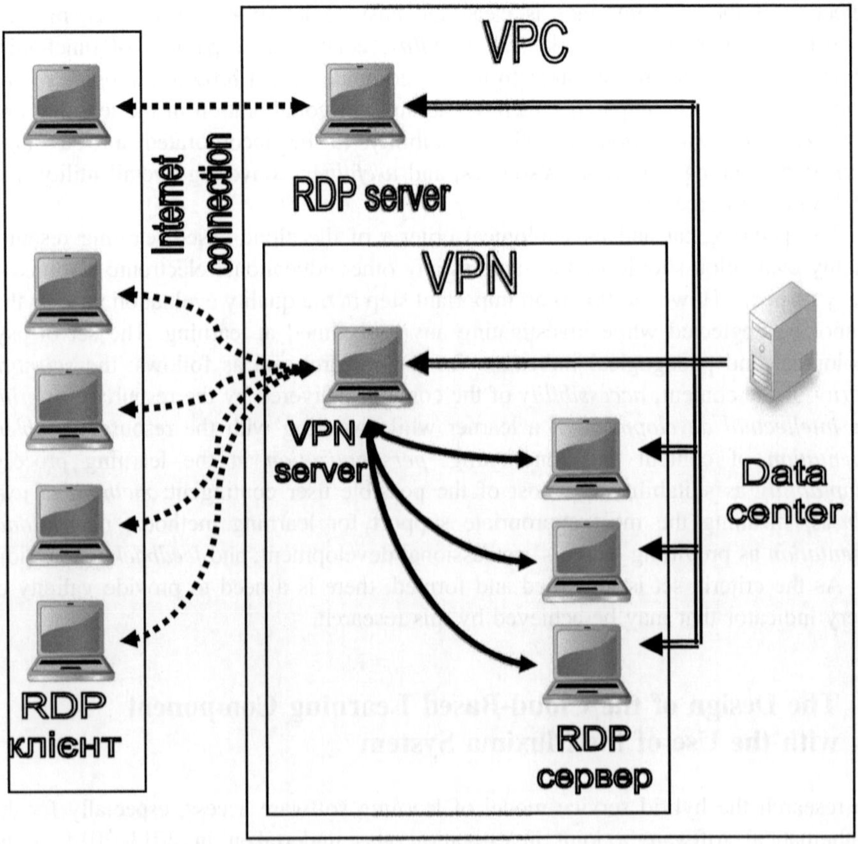

Fig. 1. The hybrid service model of the learning resources access.

In this case, a user's computer is the RDP-client, while the virtual machine in the cloud is the RDP-server. In the case of a corporate (private) subnet, a user cannot apply to the RDP-server via desktop because it is not connected to the Internet directly. Computers in the corporate subnet have Internet access via the VPN-connection, i.e. the gateway. Thus, these computers cannot be accessed from any device, but only from the specially configured one (for example, a computer in the educational institution or any other device where the VPN-connection is set up) (Fig. 1).

The advantage of the proposed model is that, in a learning process, it is necessary to use both corporate and public learning resources for special purposes. In particular, the corporate cloud contains limited access software; this may be due to the copyright being owned by an author, or the use of licensed software products, personal data and other information of corporate use. In addition, there is a considerable saving of computational resources, as the software used in the distributed mode does not require direct Internet access for each student. At the same time, there is a possibility of placing some public resources on a virtual server so the learner can access them via the Internet and use the server with the powerful processing capabilities in any place and at any time. These resources are in the public cloud and can be supplied as needed.

To create a session (to insert an item Maxima) you should choose the menu option Insert - Session - Maxima. There is an active input line to input Maxima commands. Then you can use Maxima to fulfill various operations and procedures.

An example of successful use of the Maxima cloud-based component is Graph Theory learning. Maxima has a rich set of features on the design and elaboration of relevant objects.

6 Implementation and Evaluation

In the joint research experiment held at Drohobych State Pedagogical University named after I.Franko, 240 students participated. The aim was to test the specially designed learning environment for training the operations research skills on the basis of Maxima system. During the study, the formation of students' professional competence by means of a special training method was examined. The experiment confirmed the rise of the student competence, which was shown using the χ^2 –Pearson criterion [15]. The special aspect of the study was the use of the cloud version of the Maxima system that was posted on a virtual desktop. In this case this tool was applied not only in special training situations but the students' research activity with the system extended beyond the classroom time. This, in turn, was to improve the learning outcomes.

The cloud-based learning component used in the experiment has undergone a quality estimation. The method of learning resources quality estimation developed in the joint laboratory of educational quality management with the use of ICT [7] was used and adapted for this study. The 20 experts were specially selected as having experience in teaching professional disciplines focused on the use of ICT and being involved in the evaluation process. The experts evaluated the electronic resource by two groups of parameters. The first group has contained 7 technological parameters: ease of access; the clarity of the interface; sustainability; support of collaborative work, ease of integration; mobility; and usefulness. The second group has contained 9 psychological and pedagogical parameters: the scientific clarity; accessibility; fostering the intellectual development; problem orientation; personalization; adaptability; methodical usefulness; professional orientation; and feedback connection.

Concordance coefficient W was calculated according to the formula proposed by Kendall [6] it gave the result W = 0.189 and it was significantly different from zero, so we could assume that among experts there was objective concordance. Given that the value of m(n – 1)W is distributed according to χ^2 with (n – 1) freedom degree, then

$$\chi_W^2 = \frac{12S}{m \cdot n \cdot (n+1)} = 52.8. \tag{1}$$

Comparing this value with the tabulated value χ_T^2 for $n - 1 = 15°$ of freedom and significance level $\alpha = 0{,}01$, we find $\chi_W^2 = 52{,}8 > \chi_T^2 = 30{,}5$. Therefore, the hypothesis of concordance of expert evaluations is confirmed according to Pearson.

Thus, the results of pedagogical experiment confirmed the assumption that the method of expert evaluations can be the basis for the cloud-based EER quality evaluating.

The problem was: is it reasonable and feasible to arrange the environment in a proposed way? For this purpose there were two questionnaires proposed to expert concerning two groups of parameters. The 20 experts estimated 16 parameters (there were 7 technological and 9 psychological and pedagogical among them). A four-point scale (0 (no), 1 (low), 2 (good), 3 (excellent)) was used for the questions.

The results of the evaluation of the technological parameters are shown in Fig. 2.

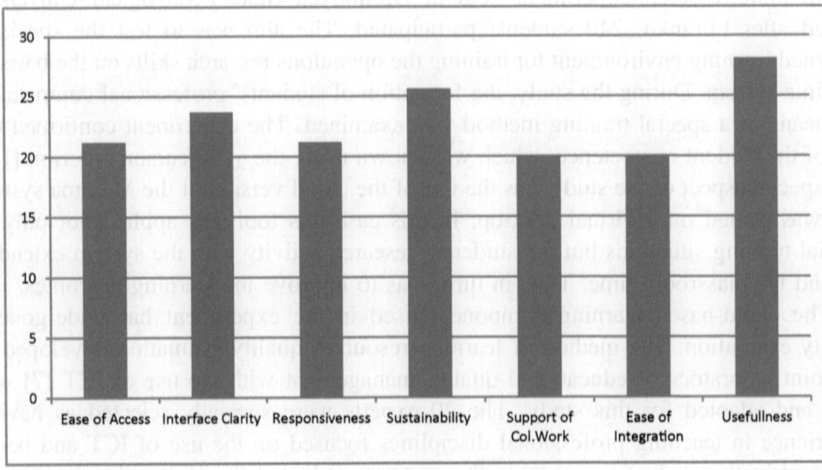

Fig. 2. The results of the cloud-based learning resource technological quality parameters evaluation.

The resulting average value was calculated for every parameter: "Ease of access" = 2.1, "Interface clarity" = 2.4, "Responsiveness" = 2.1, "Sustainability" = 2.56, "Support of Collaborative work" = 2.0, "Ease of Integration"= 2.0, "Usefulness" = 2.8, the total value was 2.3.

The results of psychological and pedagogical parameters evaluation are shown at Fig. 3.

The resulting average values for every parameter are: "Scientific clarity" = 2.6, "Accessibility" = 2.7, "Fostering the intellectual development" = 2.5, "Problem orientation" = 2.8, "Personalization"= 2.8, "Adaptability" = 2.6, "Methodical usefulness" = 2.81, "Professional orientation" = 2,75, "Feedback connection" = 2,75. The total value was 2.71.

The resulted average criterion of EER quality K = 2,59. This characterizes the resource quality as sufficient for further implementation and use.

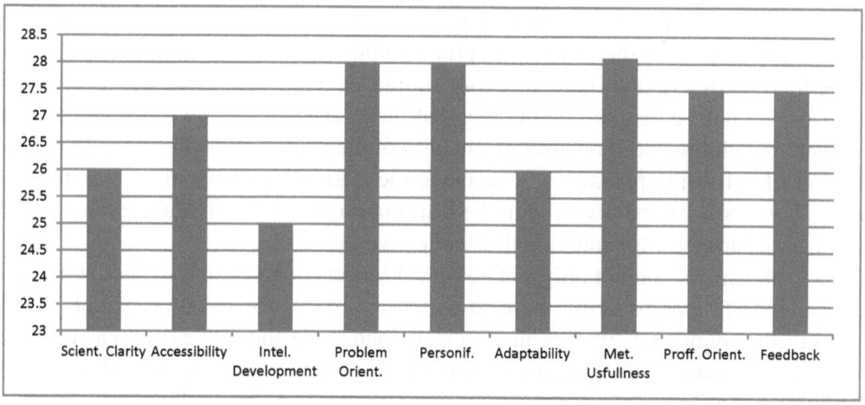

Fig. 3. The results of the psychological and pedagogical quality parameters evaluation

7 Conclusion and Discussion

The cloud services are to provide educational resources sharing, wider and more flexible access to them, in line with the principles of open education, combining of science and practice, integrating the processes of training and scientific research. The design and introduction of the cloud-based learning components into the process of Math and Computer Science study contributes to university educational environment modernization, better learning outcomes. The hybrid service model proved to be a reasonable framework to deliver and research these components. The cloud-based learning component on the basis of Maxima system was successfully used within these settings. The study showed consistency of indicators and approach to its evaluation.

Still a separate set of problems relating to the evaluation of mathematical software tools within the cloud-based settings, comparing of different service models of their delivery and pedagogical outcomes remains challenging.

By means of the cloud services the applications may become available to users being specially installed on a cloud server and provided as a public service. Recently many software applications and packages of mathematical destination have started cloud versions supplied by SaaS model that can be used in the learning process and research. In this case, access to ready-made software is supported on the vendor server. For example, Sage mathematical software is supplied by this model.

In turn, the IaaS model is designed to run any application on the cloud hardware of the provider configured and selected by the user. The IaaS composition may include hardware (servers, storage, client systems and equipment); operating systems and software (virtualization, resource management); software communication tools (net-work integration, resource management, equipment management) provided over the Internet. This raises the possibility to provide significant amount of educational content by means of quite cheap hardware (this may be a laptop, a Netbook or even a Smartphone).

Given the existence of different models of cloud services use the special attention to a balanced selection of the most appropriate solutions that fit for a particular organization, both collective and individual user should be made. The SaaS model choice in this respect could be justified by the fact that these services are the most affordable to use, even though they need a thorough analysis of the market and educationally prudent selection of software application, to achieve the desired educational or research purposes. These tools can be involved in the learning process by individual teacher, on the level of a department, for individual or collective users.

The pedagogical and not only technological aspects of software use should be taken into account while considering the type of service model for resources supply. There are many disciplines where it is necessary to outsource the processing capacity: for example, the computer design for handling vast amounts of data for graphics or video applications. This is also a useful tool used to support the collaborative work of developers, as the modern graphical applications appear to be super-powerful and require joint efforts.

For example the design of the cloud-based component with the Maxima system within the IaaS model was caused by the need to provide useful tools for graphs theory objects processing. It was necessary to support the investigative approach to learning of the Operation Research. Still this is an important problem to compare different approaches to learning settings and design taking into account various features of different software use that is not measured only by learning outcomes.

For this purpose the method of the learning components quality evaluation may be used. The advantage of the approach is the possibility to compare resources with regard to the both types of indicators such as pedagogical and also the technological ones. Due to this the learning components based on different service models and with different learning content and destination may be investigated and compared. Future research in this area should consider different types of resources and environments for Math and Computer Sciences learning from this point of view.

References

1. Bard, G.V.: Sage for Undergraduates. AMS, Providence (2015)
2. Bykov, V.: Cloud computing technologies, ICT outsourcing, and new functions of ICT Departments of Educational and Research Institutions. Inf. Technol. Educ. **10**, 8–23 (2011). (in Ukrainian)
3. Cusumano, M.: Cloud computing and SaaS as new computing platforms. Commun. ACM **53**(4), 27–29 (2010)
4. Doelitzscher, F., Sulistio, A., Reich, Ch., Kuijs, H., Wolf, D.: Private cloud for collaboration and e-learning services: from IaaS to SaaS. Computing **91**, 23–42 (2011)
5. The Future of Cloud Computing: 4th Annual Survey 2014. The North Bridge Future Of Cloud Computing Survey in Partnership with Gigaom Research (2014). http://www.bit.ly/2014FutureCloud
6. Kendall, M.: Rank Correlation Methods. Charles Griffen & Company, London (1948)

7. Kravtsov, H.M.: Methods and technologies for the quality monitoring of electronic educational resources. In: Batsakis, S., et al. (ed.) Proceedings of 11th International Conference on ICTERI 2015, Lviv, Ukraine, 14–16 May 2015, vol. 1356, pp. 311– 325. CEUR-WS.org (2015). ISSN 1613–0073. http://www.CEUR-WS.org/Vol-1356/paper_109.pdf
8. Lakshminarayanan, R., Kumar, B., Raju, M.: Cloud computing benefits for educational institutions. In: Second International Conference of the Omani Society for Educational Technology. Cornell University Library, Muscat (2013). http://www.arxiv.org/ftp/arxiv/papers/1305/1305.2616.pdf
9. Mell, P., Grance, T.: Effectively and securely using the cloud computing paradigm. NIST, Information Technology Laboratory, 10 July 2009 (2009)
10. Li, Q., Zeyuan, W., Li, W., Li, J., Wang, C., Du, R.: Applications integration in a hybrid cloud computing environment: modelling and platform. Enterpr. Inf. Syst. **7**(3), 237–271 (2013)
11. Smith, A., Bhogal, J., Sharma, M.: Cloud computing: adoption considerations for business and education. In: 2014 International Conference on Future Internet of Things and Cloud (FiCloud) (2014)
12. Shyshkina, M.: Innovative technologies for development of learning research space of educational institution, p. 1. Information Technologies and Society (2013). http://ifets.ieee.org/russian/depository/v16_i1/pdf/15.pdf. (in Russian)
13. Shyshkina, M.: Advanced technologies of e-learning in engineering education. In: Auer, M., Huba, M. (eds.) Proceedings of the 14th International Conference (ICL 2011), Piestany, Slovakia, 21–23 September 2011, pp. 565–568 (2011)
14. Shyshkina, M.: Emerging technologies for training of ICT-skilled educational personnel. In: Ermolayev, V., Mayr, H.C., Nikitchenko, M., Spivakovsky, A., Zholtkevych, G. (eds.) ICTERI 2013. CCIS, vol. 412, pp. 274–284. Springer, Cham (2013). doi:10.1007/978-3-319-03998-5_14
15. Shyshkina, M., Kohut, U.P., Bezverbnyy, I.A.: Formation of professional competence of computer science bachelors in the cloud based environment of the pedagogical university. Problems of modern teacher preparation, Uman, FOT Zhovtyy O.O., 9, part 2, pp. 136–146 (2014). (in Ukrainian)
16. Shyshkina, M.P., Spirin, O.M., Zaporozhchenko, Y.G.: Problems of informatization of education in Ukraine in the context of development of research of ICT-based tools quality estimation. J. Inf. Technol. Learn. Tools **27**(1) (2012). http://www.journal.iitta.gov.ua/index.php/itlt/article/view/632/483. (in Ukrainian)
17. Shyshkina, M.P.: The hybrid service model of electronic resources access in the cloud-based learning environment. CEUR Workshop Proc. **1356**, 295–310 (2015)
18. Turner, M., Budgen, D., Brereton, P.: Turning software into a service. Computer **36**(10), 38–44 (2003)
19. Vaquero, L.M.: EduCloud: PaaS versus IaaS cloud usage for a n advanced computer science course. IEEE Trans. Educ. **54**(4), 590–598 (2011)
20. Vouk, M.A., Rindos, A., Averitt, S.F., Bass, J., et al.: Using VCL technology to implement distributed reconfigurable data centers and computational services for educational institutions. VCL/Reconfigurable. Data Centers & Clouds/NCSU/V19-Draft 1–27 February 2009 (2009)
21. Wick, D.: Free and open-source software applications for mathematics and education. In: Proceedings of the Twenty-First Annual International Conference on Technology in Collegiate Mathematics, pp. 300–304 (2009)

22. Zhang, J., et al.: A framework of user-driven data analytics in the cloud for course management. In: Wong, S.L., et al. (eds.) Proceedings of the 18th International Conference on Computers in Education, Putrajaya, Malaysia, pp. 698–702. Asia-Pacific Society for Computers in Education (2010)
23. Zaporozhchenko, Yu., Shyshkina, M., Kravtsov, G.: Prospects of the development of the modern educational institutions' learning and research environment: to the 15th anniversary of the Institute of information technologies and learning tools of NAPS of Ukraine. Inf. Technol. Educ. **19**, 62–70 (2014)

Mobile Learning Technologies in English Learning

Nataliya Osipova, Olga Gnedkova(✉), and Denis Ushakov

Kherson State University, Kherson, Ukraine
{Natalie, Gnedkova}@ksu.ks.ua, Coldcloudu@yandex.ua

Abstract. The priority of education development is the introduction of modern information and communication technologies (ICT), in particular - technologies of mobile learning, providing improvement of educational process, accessibility and effectiveness of education for personal development according to individual inclinations, abilities, based on lifelong learning. Due to rapid expansion, proliferation and increasing of functionality of mobile technologies we should use their potential to improve and facilitate learning, ensuring accessibility, equity, individualization and flexibility. The analysis of the basic features and benefits of mobile technologies for lifelong learning is conducted. The model of mobile learning technologies in English phonetics learning is proposed in the article. At the proposed mobile learning model we developed the mobile application "English Sounds". It allows organizing the learning process of English phonetics. The methodical implementation of the proposed model for learning of English phonetics is described. The proposed mobile learning technologies help to achieve the main aim of foreign languages learning – forming the general and professional communicative competence of specialists in globalized society.

Keywords: Mobile learning · Environment · ICT · Mobile application · Pronunciation · English phonetics

1 Introduction

This article describes the development and implementation of the model of using mobile learning technologies in English phonetics learning using the mobile application, which is designed to teach practical English phonetics. Its main purpose is to increase English communicative competence and professional competence of specialists in globalized society based on lifelong learning.

As stated in the Project of National Strategy of Development of Education in Ukraine for 2012–2021 years, the aims of the National Strategy of Education are: increasing access to quality, competitive education for citizens of Ukraine in accordance with the requirements of innovative development of society, economy of the citizen; providing personal development according to the individual inclinations, abilities, needs-based on lifelong learning [1].

Mobile technologies can help in the provision of quality education of children, youth and adults around the world; it is stated in accordance the UNESCO program «Education for All» [2].

A. Ginige et al. (Eds.): ICTERI 2016, CCIS 783, pp. 169–183, 2017.
https://doi.org/10.1007/978-3-319-69965-3_10

2 Related Work

The analysis of recent researches has shown the priority of education development is introduction of modern information and communication technologies (ICT), in particular - mobile learning technologies, it ensures the educational process improvement, accessibility and effectiveness of education. Mobile learning is a new educational paradigm, so a new learning environment is created. Students have access to learning materials at anytime and anywhere, it makes the learning process comprehensive and increasing the motivation for continuing education and lifelong learning [3].

According to UNESCO, today there are more than six billion mobile phone subscribers all over the world. Due to the widespread and rapid expansion of functionality of mobile technologies, UNESCO enthusiastically uses its potential to improve and facilitate learning, accessibility, justice, individualization and flexibility for students. The European Commission is funding a number of successful R & D projects focused on mobile learning. Over the years, these projects have been gradually evolved from the develop of hardware and software to approach focused on individual, in which mobile technology plays an important role, along with a fixed technologies to support learning anytime and anywhere [2].

Mobile learning is study in or outside of classroom; it is not fixed to a particular time or place. It is supported by the use of a mobile device. Mobile devices range from standard mobile phones to tablet devices and include personal digital assistants (PDAs), MP3 players, flash drives, electronic-book readers (e-readers) and smartphones (UNESCO, 2011) [4].

«Mobile learning» , given by Geddes [5] has the possibility of mobile learning influences on behavior and intellectual of student.

Theory and practice of mobile devices usage and mobile educational resources in education are actively discussed at various scientific events. Since 2002, the International Conference «MLearnCon» is held, it focuses on applying mobile technologies in the context of learning and performance support, strategies of integrating these technologies into blended learning, and the best practices for designing, developing, and delivering mobile content [6].

In England, the conference «The Mobile Learning Network Project» (MoLeNET) is held since 2007, its aim is to spread mobile learning technologies through the implementation of the project at the same name. The result is development of a single virtual mobile learning system, gathering together about a hundred colleges and schools of the country. According to the project (2007–2010), participants explored the possibility of teaching by a variety of mobile devices: smart phones, mp3 players, tablets, voice devices, etc.

American project «Mobile Learning Environment Project» (mobile-learning environment- The MoLE) brings together participants from 22 countries who are adopting mobile learning technologies in the process of teaching in various disciplines using a single platform, connected by mobile educational resources and its methodologies.

There are interesting results of project «Mobile Technologies in Lifelong Learning: best practices» (MOTILL). The key concepts in MOTILL are Lifelong Learning and Mobile Technologies. The MOTILL project investigates how these technologies may impact on social model expansion where learning and knowledge are accessible to all, regardless of social and economic background, age, gender, religion, ethnicity or disability [7].

3 Problem Setting

Mobile learning is a part of a new pattern of education, created by technology that supports flexible, affordable, individual training. Daily use of mobile phones and other devices (tablets, digital assistants, MP3 players, flash drives, electronic-book readers and smartphones), which can be used in education, is now the main incentive of mass distribution of mobile learning.

Mobile technologies are most effective when they are integrated into a course rather than used haphazardly, so that students understand their value and see the relevance to their course [8].

The purpose of the article is design and implementation of the model of usage mobile learning technologies as a means of foreign language communicative competence formation.

Tasks of the study:

- Analyze the main features and benefits of mobile technologies for lifelong learning;
- Find the attitudes of students to use mobile learning technologies;
- Design the model of mobile learning usage in English phonetics learning;
- Development of mobile application for English articulation learning;
- Methodical implementation of mobile learning model.

4 Analysis of Main Features and Benefits of Mobile Technologies in Lifelong Learning

It is widely spread the idea of mobile learning is based on learning opportunities offered by mobile technologies. This method is most relevant when the student is not in the predefined place. Student uses situational approach and available resources. Mobile learning also allows students to easily change places, conditions and combine training in several schools. Mobile education implies the emergence of new teaching and learning methods based on cooperation in conditions of traditional classes that are often not as effective as it hoped.

Mobile learning is often called «subversive» and «changing the paradigm» , especially when its focus is shifted to education outside the traditional school or to overcome shortcomings of the existing curriculum and assessment methods. Mobile education closely connects life and learning. This type of activity is not associated exclusively with the school or university [9].

Mobile learning is applied for students' queries support, communities and social networks for learning at working place and etc. Mobile technologies allow fixing students achievement, promoting social inclusion and supporting lifelong learning [10].

Advantages of Mobile learning in lifelong learning:

- Improving of access to learning resources regardless of time or place;
- Using of inexpensive everyday technologies (hardware and software);
- Choose the own learning rhythm;
- Support of popular ways of interaction (social networks, e-mail, phone calls, SMS, forums, chat rooms);
- Fast and quality assessment and diagnosis of possible problems in learning.

The use of mobile devices is a part of modern business life, so learning directly contributes to competitiveness, the acquisition of essential skills, and working methods improvement.

Mobile learning increases student's mobility, based on unrestricted access to learning resources and flexible education. It provides global, cross-cultural possibilities.

At the adaptation of education to changing needs of students, encouraging of education continuing to renew and expand the received knowledge, creating a culture of continuous learning, students do not receive formal education, but they get to use the technologies of personal use for getting information; and the extension of their knowledge is powerful mobile learning tools to support lifelong learning education [11].

Mobile devices are used as tools for language learning. The appearance of Internet has provided open and distance learning. In a short period, the attractiveness of distance learning has led to the realization that various mobile devices provide a very effective resource for education [12].

5 Analysis of Technical and Psychological Readiness of Students to Mobile Learning

Extensive technical and functional capabilities of mobile devices in education are used as follows:

- Electronic resources, such as electronic books for mobile phones, mobile analogues of language dictionaries and reference books make possible to obtain new information regardless of time and student's place;
- Ability to play audio, graphic and video files provides the advanced capabilities, especially for learning languages. It allows the use of sources' variety and methods of obtaining knowledge and takes interest in student by unusual teaching methods;
- Internet-connection capabilities allow searching needed information, sending and receiving e-mail messages, transferring files of different files;
- Opportunity to interact with the teacher, including instant messaging to obtain advice;
- Mobile control technologies and knowledge correction allow independently monitoring the level of mastering of the material.

We conducted the questionnaire of students of Kherson State University to determine their technical and psychological readiness of mobile learning technologies usage. 160 students of the Faculty of Physics, Mathematics and Computer Science were questioned. The questionnaire showed that 99,4% of students have mobile telephones, 82% - have smartphones, 33,1% - have tablets, and 72,5% of respondents – notebooks.

Figure 1 shows quantity of percentage of students' phones, equipped with the following technical features:

1. Internet access;
2. 3G Internet access;
3. Play of MP3-files;
4. Recorder;
5. Calculator;
6. Access to Java-applications (games, e-books, etc.);
7. Camera.

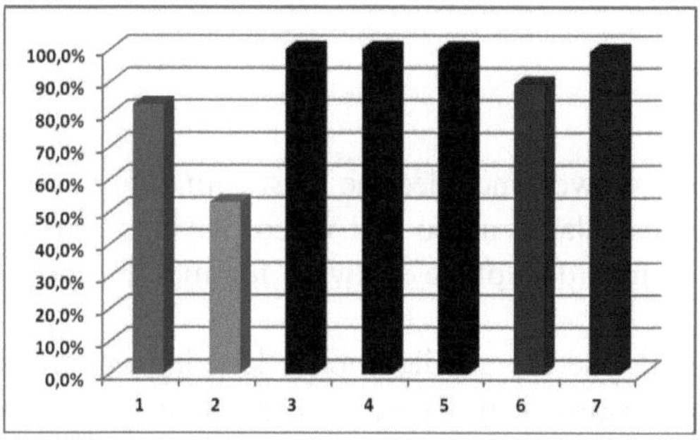

Fig. 1. The technical features of students' mobile phones

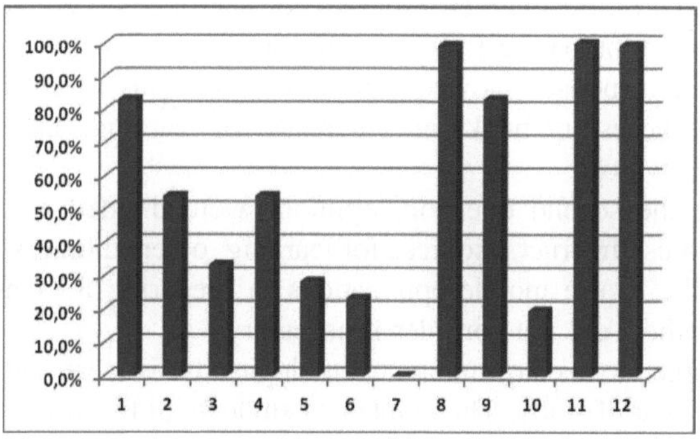

Fig. 2. Use of mobile phone applications by students

The technical features of mobile phones allow students (more than 80%) go online, use e-books, dictionaries and reference books, play audio-, video- files etc.

Figure 2 shows the applications of mobile phones are used by students, the numbers are indicated by following features:

1. Standard browser to view web pages;
2. Programs for e-mail;
3. ICQ/QIP/Jimm.
4. Programs for reading e-books;
5. Electronic dictionaries;
6. Office software (Word analogs, Excel, etc.);
7. Training programs;
8. Music player,
9. Video;
10. Voice Recorder;
11. Calculator;
12. Games.

Analyzed the answers, we concluded the most commonly used functions of mobile phones are browsers, calculator, audio and video players, e-mail. All other technical phone features (voice recorder, office software for mobile phones) are used significantly lower.

Thus, students don't use the mobile phones learning opportunities, despite rather high level of technical equipment. There is a question "Are they prepared psychologically to use mobile phones in learning? We offered the students to answer the questions:

"Would you like to download all necessary e-books and manuals for learning on your smart phone?"

"Would you like to use Internet resources for learning?"; "Would you like to use mobile applications at preparing for classes to form the necessary skills?"

Analysis of the responses showed that 82% of students would like to upload to mobile phone the necessary books and manuals for learning, and 18% of students wouldn't like to download.

At answering the second question, opinions were divided as follows: 96% of students wanted to use Internet resources for learning, others didn't want. Only 58% of students would like to use mobile applications in preparing for classes to form the necessary skills, others did not consider it necessary.

At the same time the need of mobile technologies use is increased among students, depending on the year of study. The number of students of the 4th year is more in two times the number of 1st year students who want to use mobile technology in study.

Students were asked to choose the discipline that would be convenient to study with the help of mobile technologies. More than 80% of students have chosen the foreign language. This choice is defined by the need to form foreign language skills for career development, and the opportunities offered by mobile technologies: dictionaries, electronic reference books, listening to audio files, watching videos, communicating with native speakers, etc.

Thus, the students have sufficient technical capabilities and high level of preparedness for the use of mobile technologies in education. Obviously, new possibilities of mobile learning in educational process require organizational, research and methodological work in modern policies implementation, forms and mobile learning methods.

6 Mobile Learning Model of English in Life Long Learning

According to high didactic potential of mobile learning, we designed the mobile learning model of English learning (Fig. 3).

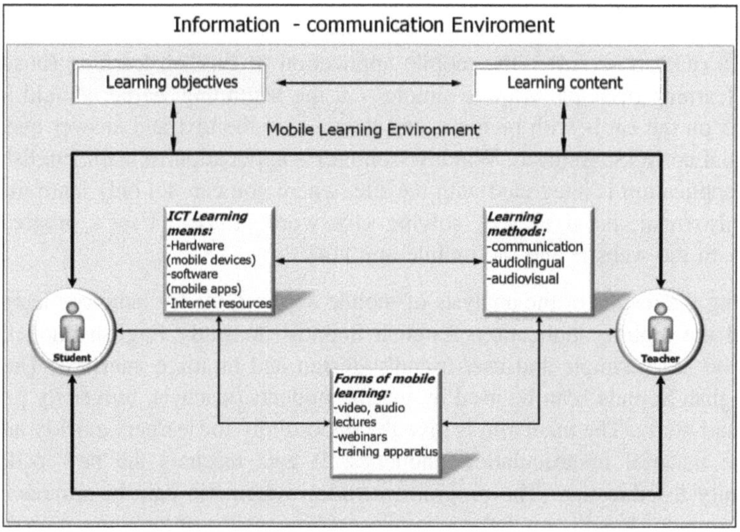

Fig. 3. Mobile learning model of English learning

Mobile learning model of foreign languages, based on the basic functions of teaching and informational-communicational technologies (ICT) opportunities; it includes a set of purposeful and orderly, and the sequence of actions of teacher and student through joint and individual study of structured learning resources [13]. Thus mobile learning in education should be based on the principle of interactive self-managed learning that will reduce destructive impact of ICT on social and cognitive learning activities. It should be noted that the most promising way of introduction of mobile technologies in education is a competent combination of new forms of learning (interactive lectures, webinars, simulations, workshops, discussions), new types of learning tasks (slide presentations, web-projects, educational podcasts) and traditional ones [7].

7 Mobile Application in English Articulation Phonetics

Mobile Assisted Language Learning (MALL) is technology of language learning using handheld mobile devices such as mobile phones (cell phones), MP3- and MP4-players, PDA, iPhone or iPad, and more. MALL is a subset of mobile learning (m-learning) and Computer Assisted Language Leaching (CALL).

Currently there are many English learning applications. Here are examples of the most popular.

1. English Grammar in Use Activities (iOS) – application, based on the well-known textbook Raymond Murphy. It is simulator that includes exercises of grammar. There are several types of exercises: choose the correct word from several proposed, write the verb in the right form, to write a response, drag words into sentences. It is possible to listen to the correct pronunciation of phrases. It is possible to view the results.
2. Busuu (iOS, Android) – the mobile application of English learning (busuu.com). The learning principle is quite simple – at the beginning learner should learn the words on the cards with pictures, and then - read the text and answer questions.
3. LinguaLeo (iOS, Android, Windows Phone) – application to learn English words. The application is integrated with the site, where you can not only learn the words, but also read; listen to text, solving crossword puzzles. User's progress syncs between the website and the mobile app [14].

Taking the results of the analysis of mobile applications for language learning, we designed the mobile application "English Sounds" to study English phonetics. The application has a simple and user-friendly design and intuitive interface. The mobile app "English Sounds" can be used by pupils, students, teachers, university professors and for self-study. The main aim is give the opportunity for learners quickly and easily learn the material in articulation phonetics. It gets teachers the new pedagogical opportunity in education. The program includes systematic data in pronunciation of English sounds. It will not only improve existing skills of pronunciation, but also improve communicative and professional competence of future specialists.

"Home page" (Fig. 4) consists of a matrix of elements - sounds. Information is presented structurally, so the division is made not only at vowels and consonants, but there is division on the subcategory (short/long, sonorous/voiceless sound, etc.). It makes the navigation and search easier and faster.

The number of icons horizontally and their size was determined based on the maximum convenience, easy and accurately pressing the buttons on the devices of different screen resolutions. The matrix is organized in such way that all its elements (vertical) fit on two screens. After clicking on the sound's icon at the main page, there is a redirect to the page with information about it (Fig. 4).

Transcription icon is presented at the top. At pressing it plays the audio by native speaker. To replay it should the click on the icon again (Fig. 5).

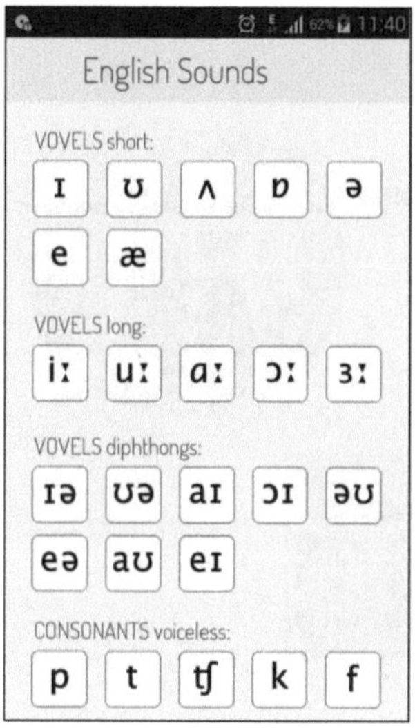

Fig. 4. Home page

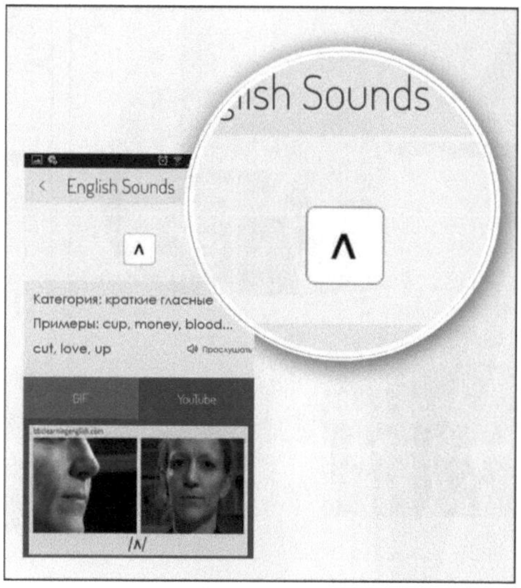

Fig. 5. Sound page

There is information about the sound belonging to one category or another, after which each of the 44 sounds are presented word-example. There is list of word-examples. At the bottom of page there are two tabs: GIF and YouTube (Figs. 6 and 7).

GIF tab is open by default and it is short (2–3 s) animated movie of pronunciation of selected audio. Material is given in two angles so that it is possible to analyze the slightest movement of vocal apparatus in pronunciation. The animation is fixated, and there is the ability to view it immediately after the opening pages with sound. To play audio does not require the Internet connection.

YouTube tab has view of the embedded video, which has word examples and comments about the correct formulation of lips and tongue at the pronunciation. The Internet connection is necessary (Wi-Fi, etc.). The application "English Sounds" is implemented with mobile devices of OS Android version 4, 5, 6. There are tools for software development: system of automatically assembly Gradle, built on Apache Ant principles and Apache Maven; IDE Android Studio; Custom TextBox with Typeface of Dosis fonts; YouTube API; component for playback SoundPool.

The mobile application has intuitive and sleek design. Color palette was chosen so that it was possible to work off the pronunciation of the sound and it would not create additional stress on eyes. The application interface automatically adapts to settings of user's mobile device interface (Fig. 8).

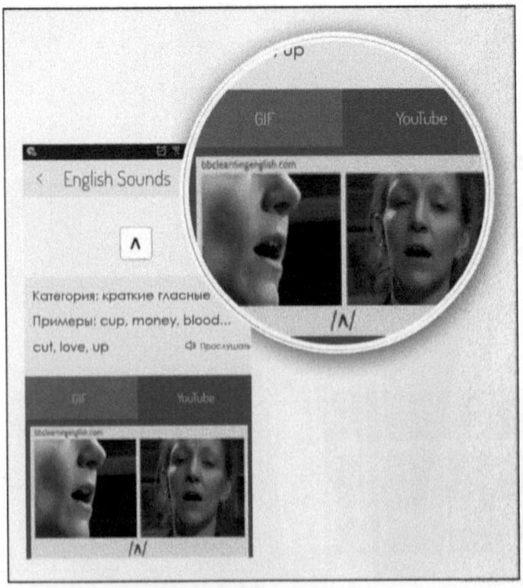

Fig. 6. GIF insert

Fig. 7. YouTube insert

Fig. 8. View of the video

8 Implementation Approaches of Mobile Application Use in Learning English

Mobile application in English learning may be used in all levels of education, in school, in high education institution, in centers of advanced training of different specialists, i.e. teachers and others.

To intensify English phonetics learning the mobile application “English Sounds” is used. It helps teacher to form and improve speaking skills and pronunciation in best way.

Discipline “English Phonetics” has the main role in English learning. It is very important to pronounce English sounds in right way to communicate. So, at forming speaking skills of students teacher should pay attention to special exercises of pronunciation of diphthongs, consonants, vowels and so on, put the stress according to type of sentences, etc.

We implemented the mobile application in distance course “Practical English Course”. “Distance learning” (DL) is the interaction between teachers and students at distance that covers all components inherent of the educational process (purpose, content, methods, organizational forms, and teaching aids) specific means of Internet technologies [15]. Thus, distance learning in language learning is not quite the same as learning of other subjects. Learning of foreign language includes not only the mastering of grammatical knowledge and comprehension skills, but interactive communication ability (competence) development [16]. With the help of distance learning system, which has all communicational and virtual tools teachers can create a virtual reality in which students can form communication skills. In teaching of language activities as reading and writing, teacher can use electronic educational resources: texts, links on article, exercises and etc. because these competences don’t require graphics, audio and video resources. However, in teaching speaking, pronunciation and listening, it is necessary to use sound, video files and illustrations to create different communication situations [17].

We developed the distance course “Practical English Course” for students of Kherson State University (Kherson, Ukraine) and for everybody who wants study English. The course is developed at distance learning system “Kherson Virtual University” (DLS “KVU”) [18]. DLS “KVU” is designed by employees of Department of Support for Academic, Informational and Communicational Infrastructure of Kherson State University. It corresponds to all requirements of international distance learning standards IMS and SCORM [19].

The aim of the course “Practical English Course” is mastering of English system and rules of its operation in different communication situations, formation of students’ linguistic and cultural skills to operate the material and acquired skills in professional activity. In the course a lot of attention is paid to study theoretical and practical phonetics. Learners study the main rules of correct pronunciation of sounds, making stress and intonation in different types of questions.

“Curriculum” contains the list of units. The unit of course includes exercises, aimed at the formation of all types of speech activity (listening, reading, writing and speaking) and tasks for self-activity, knowledge control (testing), additional learning materials (audio, video materials or online resources).

The study of vocabulary can also be accompanied by a visual annotation shown on student’s mobile devices for a better understanding of new words. In the study conducted by Chen and others, students were provided with verbal and visual annotations for learning English vocabulary. Results of testing showed that visual annotations help students with lower verbality and give more visual ability to maintain vocabulary [20].

Listening skills can be considered the first stage in language learning. With development of second generation of mobile phones, it has become possible to develop a mobile multimedia system for learning listening skills through audio exercises [21].

In distance course the exercises in listening are developed (Fig. 9). Students listen to the audio file and do the exercise.

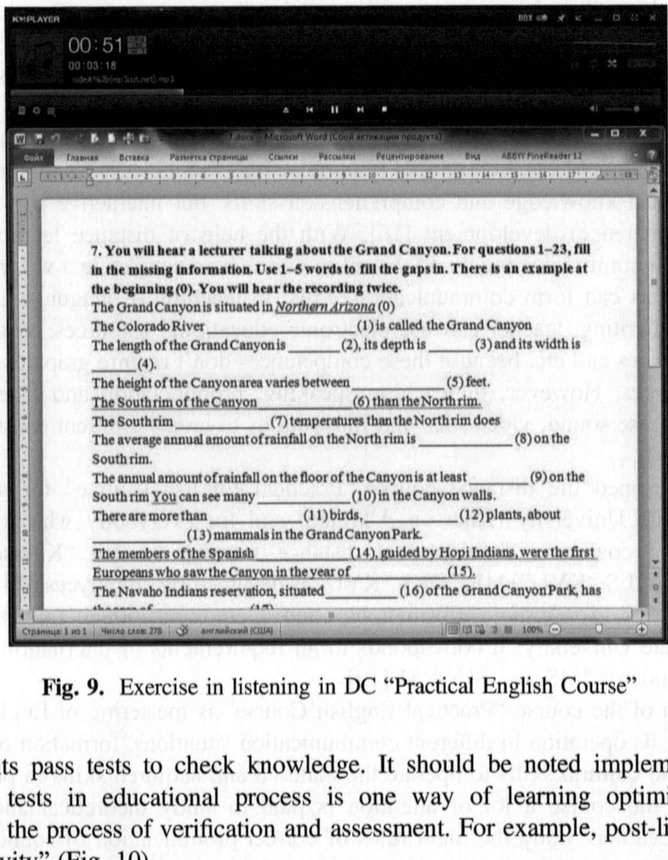

Fig. 9. Exercise in listening in DC "Practical English Course"

Students pass tests to check knowledge. It should be noted implementation of electronic tests in educational process is one way of learning optimization and improving the process of verification and assessment. For example, post-listening test "Associativity" (Fig. 10).

So, mobile devices allow organize language learning process at any time and any place during all life.

There are exercises in phonetics study. We applied the mobile application "English Sounds" in phonetics learning in the distance course. For example, students should study the sounds and listen to the pronunciation and repeat it. The exercises student should do at home or in class. The exercise is presented in Fig. 11.

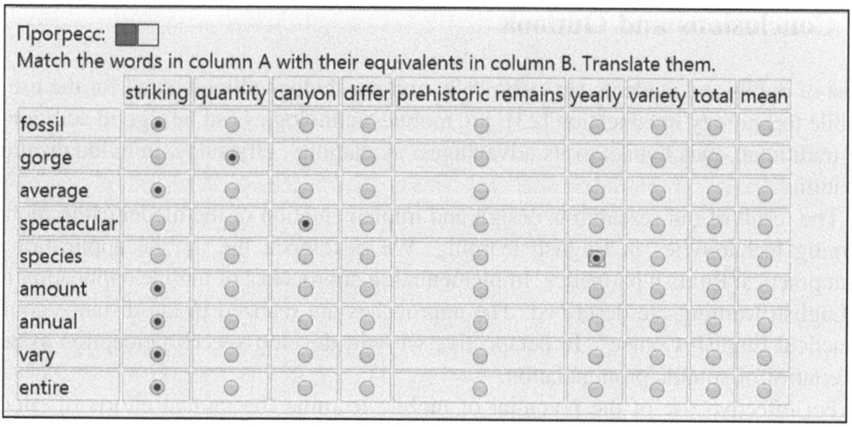

Fig. 10. Test question "Associativity"

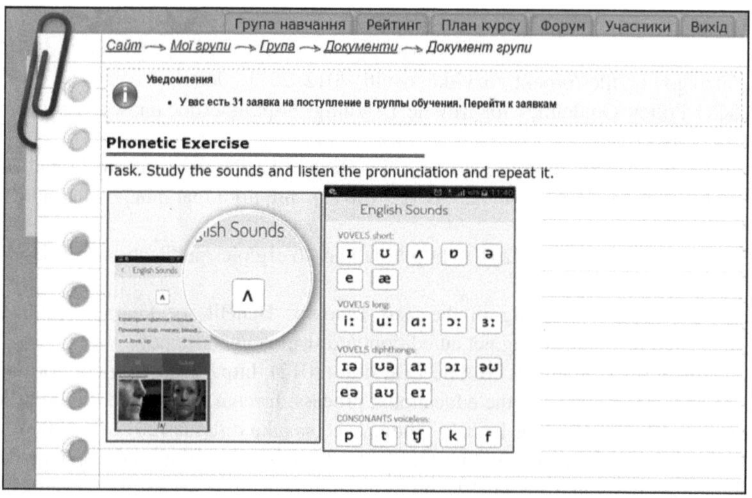

Fig. 11. Exercise in Phonetics using mobile application "English Sounds"

So, usage of mobile technologies in learning process improves all types of communication abilities, especially speaking and pronunciation skills. Communication has the great role in mobile learning. Different learners are able exchange their knowledge, skills and attitudes through interaction in some virtual educational environment. It is called collaborative learning. Collaborative learning helps the learners to support, motivate and evaluate each other to achieve substantial amounts of learning, the property which is almost absent in other kinds of learning. One can attain a good collaborative approach simply by using a mobile device as an environment for learning, which is highly dependent of the users than the devices [22].

9 Conclusions and Outlook

Most of pupils and students are technically and psychologically prepared for the use of mobile technology in education [23]. So, mobile technologies can be a good addition to the traditional ones thanks to its advantages: availability, efficiency, individualization, flexibility.

The result of our research is design and implementation of the model using mobile learning technologies in English learning. We developed the mobile application to learn practical English phonetics. Implementation approaches of mobile application use in English learning are described. The approaches are realized in the distance course "Practical English Course". In perspective we will develop speech perception system for control of sounds pronunciation.

For effective use of the potential of mobile learning the mutual efforts of educational leaders, software developers, trainers and teachers are required.

References

1. Project of National Strategy for the Development of Education in Ukraine for 2012–2021. http://iitzo.gov.ua/files/proekt_rozvitku_osviti_2012_2021_.doc
2. UNESCO Policy Guidelines for Mobile Learning. http://unesdoc.unesco.org/images/0021/002196/219641E.pdf
3. Rashevska, N.: Mobile software training. In: Rashevska, N. (ed.) Informational Technologies and Learning Tools, vol. 1, no. 21 (2011). http://journal.iitta.gov.ua/index.php/itlt/article/view/369
4. UNESCO. Mobile Learning (2010). http://iite.unesco.org/pics/publications/ru/files/3214679.pdf
5. Geddes, S.: Mobile Learning in the 21st Century: Benefit for Learners (2004). http://knowledgetree.flexiblelearning.net.au/edition06/download/geddes.pdf
6. 13th International Conference Mobile Learning (2012). http://mlearning-conf.org
7. Fine, M.: Mobile learning in the educational process: foreign experience. Modern Scientific Research and Innovation, № 1 (2015). http://web.snauka.ru/issues/2015/01/43006
8. Edirisingha, P., Salmon, G., Fothergill, J.: Profcasting – a pilot study and guidance for integrating podcasts in a blended learning environment. In: Bernath, U., Sangrà, A. (eds.) Research on Competence Development in Online Distance Education and E-Learning, pp. 127–137. BIS-Verlag, Oldenburg (2007)
9. http://iite.unesco.org/pics/publications/ru/files/3214679.pdf
10. MOTILL – Mobile technologies in lifelong learning: recommendations. http://www.motill.eu
11. United Nations Educational, Scientific and Cultural Organization. www.unesco.org/
12. Kukulska-Hulme, A., Shield, L.: An overview of mobile assisted language learning: can mobile devices support collaborative practice in speaking and listening? http://portal.acm.org/citation.cfm?id=1520087. Accessed July 2009
13. Gnedkova, O., Lyakutin, V.: Mobile learning model designing in distance learning system "kherson virtual university". Inf. Technol. Educ. **24**, 107–118 (2015). (In Ukrainian)
14. Guide to "apps": 5 applications for learning English. Arguments and facts. http://www.aif.ru/tech_no/article/51718

15. Polat, E.S., Buharkina, M.U., Moiseeva, M.V. (eds.): Theory and Practice of Distance Education: Educational Material for the Students of High Pedagogical Educational Institutions, p. 416. Publishing centre "Academia", Moscow (2004)
16. Kravtsov, H., Gnedkova, O.: The use of cloud services for learning foreign language (English). In: Ermolayev, V., et al. (eds.) Proceedings of 12th International Conference on ICTERI 2016, Kyiv, Ukraine, June 21–24, pp. 620–631. CEUR-WS.org/Vol-1614 (2016). ISSN 1613-0073. http://ceur-ws.org/Vol-1614/paper_87.pdf
17. Olijnik, O.: Innovative technologies of distance learning a foreign language for students of non-language universities. Linguistic research: Research works of HNPU named after Skovoroda. - 2014 – Ser.38 (2014)
18. LMS. Kherson Virtual University. http://dls.ksu.kherson.ua
19. Kravtsov, H., Kravtsov, D.: Knowledge control model of distance learning system on IMS standard. In: Iskander, M. (ed.) Innovative Techniques in Instruction Technology, E-learning, E-assessment, and Education. Springer, Dordrecht (2008). doi:10.1007/978-1-4020-8739-4_35
20. Chen, C.M., Hsu, S.-H.: Personalized intelligent mobile learning system for supporting effective English learning. Educ. Technol. Soc. **11**(3), 153–180 (2008)
21. Huang, C., Sun, P.: Using mobile technologies to support mobile multimedia English listening exercises in daily life. In: The International Conference on Computer and Network Technologies in Education (CNTE 2010) (2010)
22. Yang Soon, G.: Technology for Foreign Language Learning (2005). http://www.readingmatrix.com/conference/pp/proceedings/gohyingsoon.pdf. Accessed Dec 2010
23. Spivakovska, E., Osipova, N., Vinnik, M., Tarasich, Y.: Information competence of university students in Ukraine: development status and prospects. In: Ermolayev, V., Mayr, H.C., Nikitchenko, M., Spivakovsky, A., Zholtkevych, G. (eds.) Information and Communication Technologies in Education, Research and Industrial Applications. CCIS, vol. 469, pp. 194–216. Springer, Heidelberg (2014). doi:10.1007/978-3-319-13206-8_10

The Experience of the Master Classes as a Means of Formation of Readiness of Teachers to Implement Innovation

Nataliya Kushnir[1(✉)], Nataliya Osipova[1], Nataliya Valko[1], and Olena Litvinenko[2]

[1] Kherson State University, Kherson, Ukraine
{kushnir,natalie,valko}@ksu.ks.ua
[2] Kherson National Technical University, Kherson, Ukraine
mmkntu@gmail.com

Abstract. The article describes the experience of the master classes for school teachers and university lecturers to use ICT in the educational process. Improving teacher's/lecturer's qualification involves their willingness to follow the advanced technology. However, there are several factors that should be considered in the preparation of training courses for teachers. The authors consider the technology of the master classes as an alternative to long-term courses. This approach has several advantages: short term, productivity at work. During the master-class the main goal is not to teach how to use the technology perfectly but to show its advantages, persuade to use a particular technology, demonstrate the ease of use, and motivate teachers for the further use of the offered technologies in their professional activity. The experience of master classes for teachers of Kherson and Kherson region over the past 5 years was analyzed.

Keywords: Teaching strategies · ICT tool · Teaching methodology · Capability · Technology

1 Introduction

This article describes the purpose and methodology of the master classes for the successful teachers' training programs on the use of appropriate ICTs and their knowledge of the trends and new approaches in the education system.

The introduction of new educational technologies in the educational process depends primarily on the willingness of a teacher. Without proper motivation of the teacher the best teaching ideas will not be implemented simply by the order. The famous Pareto principle 20/80 can be interpreted as following: only 20% of teachers supports introducing innovations, 12% of them are passive ones, and 8% are active ones who encourage, develop and improve the technology. The majority of teachers (80%) oppose any innovation. This position does not depend on the innovation itself. However, trends in society require the restructuring of the education system and active introduction of innovation, which are able to prepare students for a successful life in the information society.

A. Ginige et al. (Eds.): ICTERI 2016, CCIS 783, pp. 184–199, 2017.
https://doi.org/10.1007/978-3-319-69965-3_11

2 Related Work

Programs of involving teachers in learning using ICT are implemented by many researchers. At the same time general researches are conducted, for example, associated with the formation of teachers' ICT competence and aimed at effective implementation in the educational process and the further development of individual approaches (mobile studying, use of social networks in learning process and so on). Thus, in the research [10] a mobile learning project and experience in the use of mobile technologies in the training of teachers is described. The creation of flexible learning solutions that will allow teachers to have access to information using a variety of devices and training support in different situations is shown in the paper.

Another perspective direction of development of education is using social networks in learning [1]. Practical activities in the field of education, i.e. using of social networks in work to support learning, interaction of teachers and students, creating business relationships and knowledge sharing.

Analysis of the studies shows that the use of cloud technology is already normal practice for one third of population. In the future, this number will only increase and leading position will be owned by those companies and organizations that will offer their services using cloud resources, like services that can be customized to individual liking. This also applies to educational technology. Firstly cloud technologies allow universal access to data at any time. Secondly, they contribute to the construction of their own learning trajectory at the right pace.

In paper [16] the authors note several advantages of using cloud technologies, namely:

- Users have access to a wide variety of instructional and informational resources – not just MOOCs (Massive Open Online Courses) but videos, podcasts, and other resources.
- Users build a professional network of trusted colleagues – both inside and outside their organization – with whom they communicate to exchange ideas, resources, and study sometimes without realizing it. Many refer to their network as their Personal Learning Network (or PLN).
- Users are keeping up to date with what's happening in their industry and profession use such ways as reading blogs and web feeds and pages of internet-journals.

According to research [4] 50% of the adult population of Ukraine (over 15) use online resources, and two-thirds of them use social networking services. Modern gadgets (smartphones, tablets, laptops) are owned by 57% of the population.

Conducted by UNESCO [15] research has shown that without appropriate training educators use technology only to solve their former tasks. In this case there is no transformation and qualitative growth of the level of teaching and learning.

As the recommendations UNESCO offers [15] the following stages of the introduction of ICT in educational practice:

- Demonstrate and model how mobile technologies can improve the quality of teaching, learning and management of the educational process.

- Share research data and analysis of mobile learning programs.
- Encourage dialogue between the main participants, including principals, teachers, students, parents, local authorities and the community on mobile technologies training.
- Provide a consistent concept of how technology, including mobile one, can contribute to the achievement of learning objectives.

However, there are some limitations such as the lack of theoretical and pedagogical foundations, sustainable integration in formal educational contexts and, in particular, the lack of support for teachers and training [2].

3 Problem Setting

Learning each trend is time-consuming process for a teacher. However, it is important to form teachers' sustained interest to the changes in society and the educational system under the influence of digital technologies. The main factors that need to convince teachers are:

- Change the role of ICT in Education (shift from a focus on "Education for ICT" to the use of "ICT for education" [16])
- Use of digital devices which are natural part of the life of modern students' generation that meet their educational expectations.
- Emergence of new approaches in education such as SMART, IoT, STEM education (Science Technology Engineering Math), BigData, MOOC, BYOD (Bring Your Own Device), Blended Learning, Flipped Classroom, mobile learning, cloud, gamification of education, etc.
- Change in the spectrum of required specialties on a labor market (now and in the future) and the need for learning throughout life.

Thus, there is a need not only to teach the teachers to use certain software products and services in the work, but to form an understanding of the society development trends and their own willingness to learn new technology. Since 2012 we research the subject of teachers' training in ICT, use new approaches in teaching that meet the educational requirements of the today's youth, including training teachers through master classes [8]. In our view, conducting master classes are the most effective form for solving these tasks.

The purpose of the article is to describe the experience of the master classes as a means of formation of readiness of teachers to implement innovation.

Tasks:

- to analyze the main trends in modern education;
- to reveal teachers' attitude to the use of ICT in educational process;
- to determine the understanding of general trends in the development of education under the influence of ICT;
- to conduct master classes with the use of ICT tools and services for teachers.

In our view, for effective implementation of innovation each teacher should take certain steps:

1. Get knowledge about the nature, advantages and disadvantages of a particular innovation.
2. Accept innovation as the integral part.
3. Develop necessary competences, in particular informational.
4. Start using innovation.
5. Develop and spread innovation.

There are both objective and subjective factors that explain the reluctance of most teachers to implement innovation:

- Mastering of innovation requires teacher to spend more time, which is not paid.
- It is necessary to rebuild style of work and usual teaching methods that causes a situation of discomfort.

Analyzing the results of the anonymous surveys, interviews with teachers, we find an explanation which is: "innovation – is just a passing fad", "the Headmaster wants to gain an authority", "I do not have opportunities to master innovation due to lack of free time" and others.

The article [9] considered three levels of information competence (I level – beginner; II level – active user; III level – expert). Based on these levels the authors propose to divide teachers' information competence according to their ability and willingness to use ICT in the learning process. The article also identifies the major reasons of teachers' unpreparedness for using ICT in learning. They include lack of motivation for using ICT; lack of complexity; learning computer skills only without the support of innovative educational technologies; ignoring the characteristics of adult education; neglect of interactive teaching methods; insufficient integration of knowledge and skills of students from different academic disciplines; insufficient formation of Computer Science teachers the skill concept of the 21st century.

Today in our country there is a situation when teachers are supported by large companies and famous brands for the production of computer hardware and software. Since 2005 in Ukraine education programs were adopted: "Intel® Learning for the Future" and the educational program "Partnership in Learning", "Teachers on-line" and "Expert Teacher" (Microsoft). Also recently the Ministry of Education of Ukraine announced the cooperation with the company LEGO in the implementation of the robotics into learning process.

Learning is available always (available internet), everywhere (mobile phones), unlimited computing capacity (instruments).

For the study there are available not only text files with pictures. There are many educational channels (video content), educational online academies (video, text, animation, online laboratories), forums and discussions. Cloud technology allows organizing joint activities, teaching. Learning can take place at any rate, for any type of perception. Information for learning is more than enough.

But to get success we need not only technologies. The learning process is more efficient in interaction with person. Therefore, in Ukraine there is practice of holding master classes among educators. These master classes are held under the protection of well-known companies (Microsoft, Google, Intel), providing information on new products and services and sharing the experience of successful use of the software. These master classes are conducted with the efforts of teachers, (those who support and spread) of current trends (Artur Kocharyan, Margarita Kalyuzhna, Ivan Shyhat-Sarkisov and Vladislav Breger).

The school of programming are created (STEP, Vinahidnik), and free, on voluntary basis (IT2School, Technic.in.ua) In support of these clubs different communities were created, such as volunteer network of coding (http://codeclubua.org), the educational project Microsoft (https://www.microsoft.com/uk-ua/education), GoogleAcademy (http://scholar.google.com.ua) and others.

Under the impact of these initiatives teachers create their own educational spaces using ICT tools. Many teachers and lecturers are the professional bloggers, websites owners. An example is a community founded by Ivan Ivanov "ICT training of teachers in Ukraine" and its resource "To be able to live. The development of the XXI century skills using digital technology". This project was responded by hundreds of teachers and professors.

Teachers begin to link their professional growth not with deepen and broaden the knowledge of the school subject, but with the improvement of general pedagogical skills and knowledge. A special role in the professional growth of teachers belongs to the pedagogical aspects of ICT use.

Increasing the role of ICT in the educational process contributes to changing pedagogical techniques.

4 History of Development of Master Class System in KSU

Conducting master classes in Kherson State University was founded in 2006 under the Intel program "Teaching for the Future." As a result of questioning teachers of pedagogical universities, who were trained under the program, expressed a desire to use IT in their work and realized that IT will develop an educational approach that focuses on the students' needs.

The cooperation of the Ministry of Education and Science of Ukraine with the Microsoft implementation of the joint program "Partnership in Learning" (Memorandum of Understanding of 28 October 2003. Protocol № 1 of 3 August 2004), to the order of MES # 693 of December 6, 2005. a program to train teachers and students of pedagogical universities 'Basis of Information Technologies' is implemented. Within the cooperation with pedagogical institutions of I–IV levels "Partnership in Learning' program supports 36 h of seminars "Basics of Information Technology" for teachers of high schools. In 2009, 2 seminars for teachers of KSU were held. 17 university teachers took part in it.

The next step was the introduction of the course "Use of information technologies in educational process" for students of specialties of Physics, Mathematics, and a Primary School Teacher. However, this course is taught only to those teachers of with Computer Science specialization. At the same time, we saw that the students are sometimes hard to see the potential of a particular technology to be used in the educational process. They lacked the practice of working with school students. We developed the idea of organizing joint work of school teachers and students. We have chosen the master class as the best form. It is a modern form of educational training seminar for practical skills on different techniques and technologies to improve the professional level and share best experiences of participants, expanding horizons and introduction to the newest branches of knowledge.

Since 2012, we started cooperating with a number of Kherson schools and conducted master classes for teachers of educational institutions based on them. A feature of these master classes was that the teachers worked with students – future teachers. Students, as representatives of Net generation, easily mastered the technology, and teachers have seen better educational opportunities of new service [13]. Also, during the master classes we use gaming technology, brainstorming, group work, pair work. More detail on this experience in the article in 2014 [7], where the focusing was on the benefits of this technology for students.

Joint work of teachers and students in the format that often hard to implement. This is due to the time issue for every participant: students, teachers, professors [14]. However, today the need for school teacher training is sharply increasing, and that's where the system of master classes can be an effective tool.

5 Setting up the Pedagogical Experiment

According to the importance of the formation of teachers understanding of education development trends under the influence of digital technologies, their willingness to self-development and use of new techniques and services we found the possibility of holding master classes for teachers of Kherson region who are passing courses on the basis of Kherson Academy of Continuous Education. So, master class "Modern ICT tools and services for teachers of Mathematical and Technology disciplines" was first performed by the teachers of the department of Computer Science, Software Engineering and Economic Cybernetics of Kherson State University as part of the All-Ukrainian scientific-practical web conference with international participation "Innovative Dimension of Development of Mathematics and Technology Education", held in October 29–30, 2015. This master class was held for 4 groups of Physics, Mathematics and Computer Science teachers in Kherson region.

Interest in the master classes was shown by the teachers of the Department of Physics and Methods of teaching of the Faculty of Physics, Mathematics and Computer Science in Kherson State University. The effectiveness of the master class significantly was improved through joint participation of school teachers and university professors. It made possible to pay attention to the peculiarities of the implementation of innovative approaches using ICT services in accordance with the methods of teaching school Physics course.

During the master classes our objective was to present teachers common tools that they can use in future professional activity in school regardless of the subject, to develop participants' interest to learning ICT and pedagogical and technological capabilities of the services, the ability and willingness to use them in their professional activities, to create an electronic bank of ideas on the use of ICT services for the solution of pedagogical problems.

The master class program includes the following issues:

1. Consideration of conflicts arising due to psychological characteristics of students with using modern and traditional approaches, methods and means of training.
2. Introduction to the concept of teachers and new online services for education:
   - 'Flipped classroom'; highlighting educational opportunities and services (YouTube and International Educational Network of Khan Academy).
   - Use of Content management systems (CMS) and Learning Management Systems (LMS) (as examples there are systems used in Kherson State University: KSUOnline [6] and Virtual Kherson University [5]).
   - Create video tutorials for using services EdPuzzle and Blendspace, allowing to search desired videos from popular video sharing like YouTube, Vimeo, Khan Academy, LearnZillion etc. Create exercises, quizzes, additional questions and comments. Services include creating classes and have powerful features statistics.
   - Development of educational games using the service Learningapps.
   - Presentation to Internet services for the research and design activity for the Concept of STEM-education (Interactive works of virtual educational laboratory www.virtulab.net allow to conduct virtual experiments in Physics, Chemistry, Biology, Ecology and other subjects, in two-dimensional and in three-dimensional as well. Globallab – a platform that serves as the international environment of research interaction between students).
   - Introduction to interactive services and chronicles creation Dipity and Tiki-Toki.
   - Using of network services for the organization of modern educational process.
3. At modeling stage teachers were suggested to work in groups and to develop a lesson using Internet services: interactive video tutorial with test tasks; training game with the chosen theme; representing chronicle of historical facts related to the chosen training topic; publishing links to educational materials created in chosen CMS, LMS or blog.
4. Demonstration and discussion of copyright material for the lesson with the master class participants, use of these tools for enhancing learning activities.
5. At the stage of reflection there is a discussion on the results of joint activities; a survey of teachers to identify their satisfaction with the results of master class and intentions on how to use learnt technology in teaching is conducted.

Various online services use in their basis the principles of teamwork, cooperation, openness and accessibility. When choosing a service for the master class, we were guided by the following criteria:

- The feasibility
- Easiness to learn and to use the service
- Interactivity

The successful mastering of the master class theme was based on the productive activities of all participants.

During master classes these materials were used: "Introduction to Information Technology" and "Master Class" published on the website http://ksuonline.kspu.edu/.

Implementation scenarios and approaches, as well as specific details of the master classes' execution, which could all have been useful for those willing to replicate the approach was published earlier [7].

6 Survey Analysis

For the final phase of the master class a survey was prepared. The survey contained 19 questions of closed type in three blocks:

- general information;
- experience in using ICT in the classroom;
- tendencies of changes in the education system.

The first set of questions concerned the general information about work experience, school location, and classes the teacher works with.

Among the interrogated there were many teachers with great work experience: 88% have experience of more than 10 years. Majority of teachers work in secondary and high school (1–4 grades – 12%, 5–11 grades – 88%) (Fig. 1).

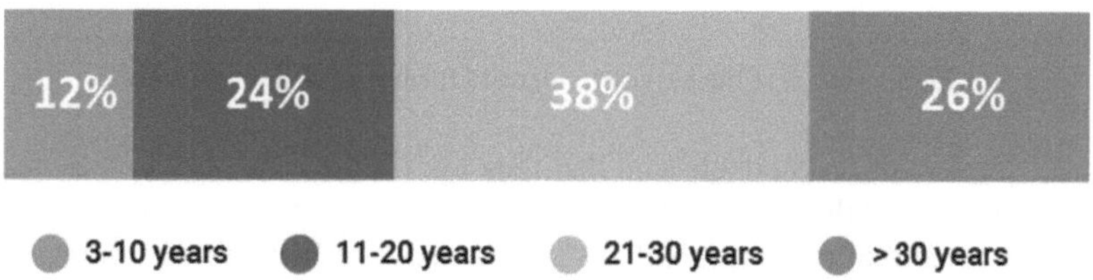

Fig. 1. Representation of teachers regarding work experience

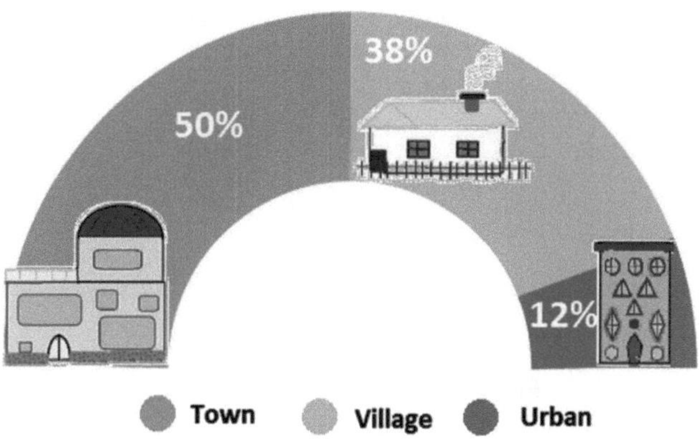

Fig. 2. Representation of teachers regarding school's location

As a master-class was conducted for teachers of training courses, most of the participants came from nearby towns and villages (88%) (Fig. 2).

The second set of questions concerned the identification of experiences and purposes of using ICT in professional work. Questions contained answers to the multi-selection (checkbox) and the line to write additional information (Fig. 3).

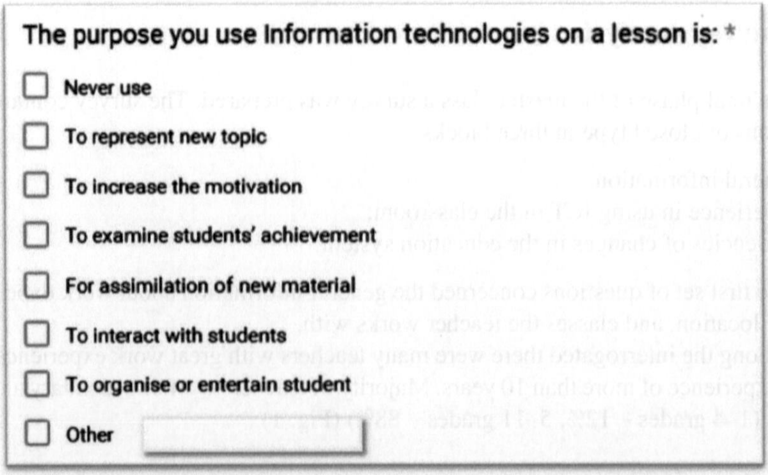

Fig. 3. Detail of the survey to identify the use of ICT experience

Three-quarters of respondents said that they use different devices in their practice, and a quarter said that they were going to use them, but do not have this opportunity yet. It should be noted that some respondents (3%) noted that their schools require lessons using ICT: teachers have to conduct lessons with an interactive whiteboard in the computer lab or using other ICT (Fig. 4).

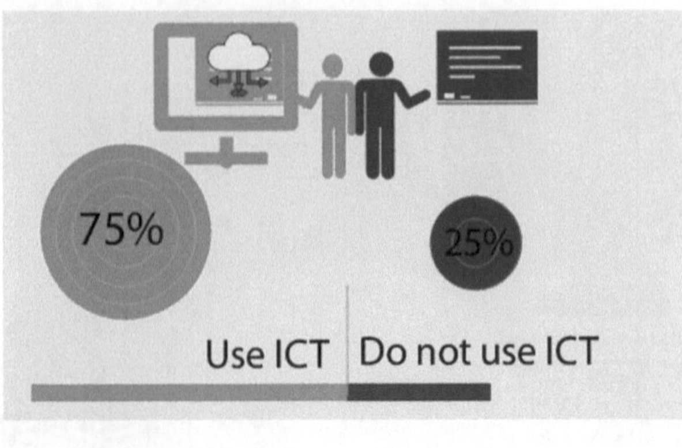

Fig. 4. Use ICT in teachers practice

Analyzing responses of 75% of teachers who use ICT, we found out the purpose of using ICT and distinguished the most popular activities. Traditionally, the most common purpose is to explain new material – 28% and conducting current and final examination – 25% (Table 1). 18% of teachers use ICT in order to increase motivation and interest of students. Another 17% of respondents use ICT to assimilate the material. None of the teachers gave their own version. The most difficult activity appeared the organization of communication and collaboration among students using digital technologies. From our point of view, this is due to the difficulty for the teacher to assess the contribution of each student in the joint work, the lack of formation of the teachers skills of cooperation organization of students work in groups.

Table 1. Activities with using of ICT

Activity	Answers, %
Explaining new material	28%
Increasing motivation	18%
Conducting current of final examination	25%
Assimilation of new material	17%
Improving communication with students	4%
Organization or entertaining of students	8%
Other	0%

On the question about their experience in using ICT in the classroom the following answers were received: laptop/ computer (100%), video (87%), television (37%), tablet (25%), an interactive whiteboard (12%) (Fig. 5). Some respondents chose more than one

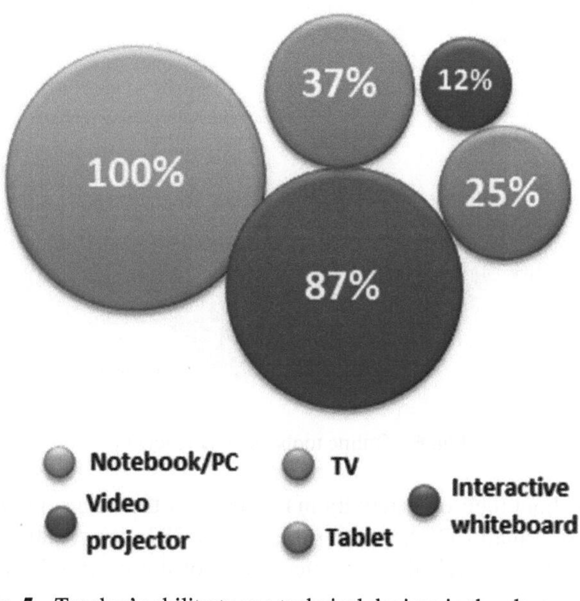

Fig. 5. Teacher's ability to use technical devices in the classroom

answer. These data have improved significantly compared with surveys of previous years. They confirm the understanding of the need to use digital technologies in the classroom. However, the technical equipment is not always used effectively. It should also be noted that we do not show the statistics of equipped classrooms in the city and region where the average numbers may be very different. We present results of a survey conducted in November 2015 and January 2016 and November 2016 among teachers who held a master class. The findings of our survey data are correlated with the survey conducted among teachers of Computer Science in February 2015 [11].

However, the teacher said that the possibility of using digital devices by students is less. Thus, almost all students are able to use computers in the classroom, 12% of teachers give students an opportunity to use interactive whiteboards and mobile phones. Half of teachers (50%) indicated that students have the opportunity to use tablet on the lesson.

Another key question was: “What tools of ICT do you use ICT to organize the work and work of students?”. This question has been open type (had to write their own tools), so we got a lot of answers. All responses were grouped into the following categories:

- Social networking/ communications (e-mail, Skype, facebook).
- Blogs/ sites (blogger, Sites).
- Educational platforms (Moodle, Microsoft educator network).
- Cloud storage (GoogleDrive, OneDrive).
- Office applications (MSOffice, Office 365).
- Online tools (Mobile Apps Google Apps, Google-Services, Youtube, Movie Maker, Cacoo, FreeMind, Wunderlist, ToDoList, Padlet, Popplet, Conceptboard etc.)

Survey results (Fig. 6).

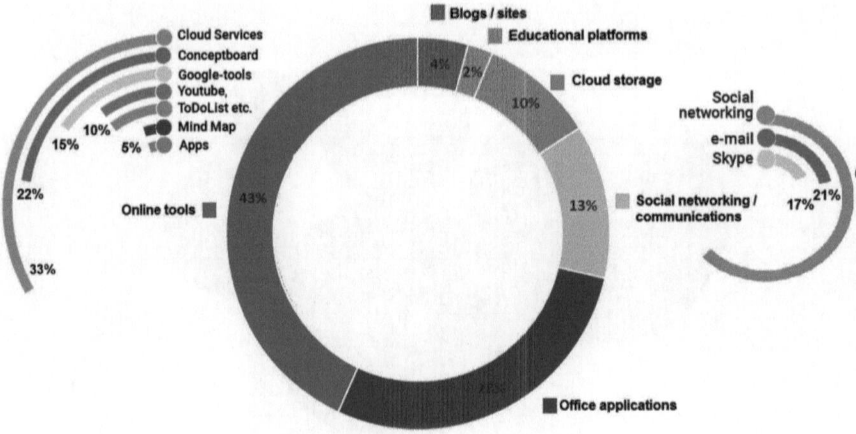

Fig. 6. Online tools used by teachers

When asked about their own activity in the network the largest number of respondents said they use e-mail (33%). Among these, only 28% use social networking, 17% have blogs, and 11% – a website. The intersection of these activities as blogging and maintaining the site, was not observed in the responses. Among the responses were others, including the absence of any activity (11%) (Fig. 7).

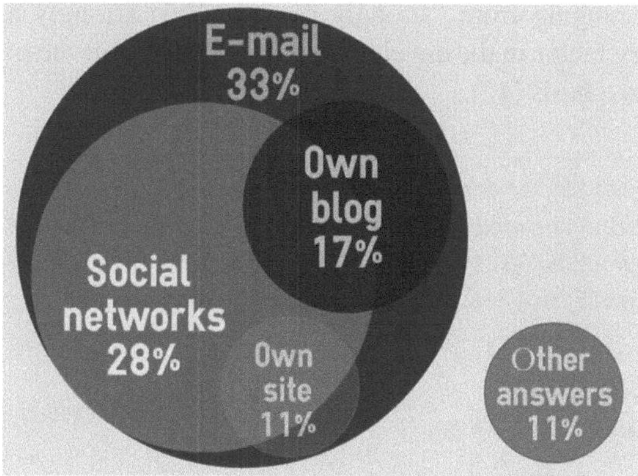

Fig. 7. Services used by teachers

One of the tasks we have set is to show teachers the opportunity to create their own information space. They can use a website or blog. Comparing the results of the answer to this question, we asked teachers in 2012, it should be noted that the percentage of teachers using websites and blogs in their work has increased. So, we can note the increase of use of blogs from 5% to 17%, sites – from 2% to 11%. E-mail is started using by 7% more. The number of social network users didn't change.

As a result of participation of master class 100% of teachers have created e-mail account, which remained active 2 months later. All existing sites and blogs for master class were supplemented by links to resources (mostly YouTube, EDpuzzle, LearningApps) (Fig. 8).

Fig. 8. Example mathematics teacher's Blog (Inna Prydius)

The last set of questions was developed with desirable psychometric properties to measure (detect) the level of awareness of current trends of changes in the education system (Fig. 9). This block contains 6 questions and it was developed using a four-point Likert-type scale, ranging from "strongly disagree" to "strongly agree". The trends highlighted as a key factor in the development of education for the period from 2010 to 2015 were taken as a basis [12].

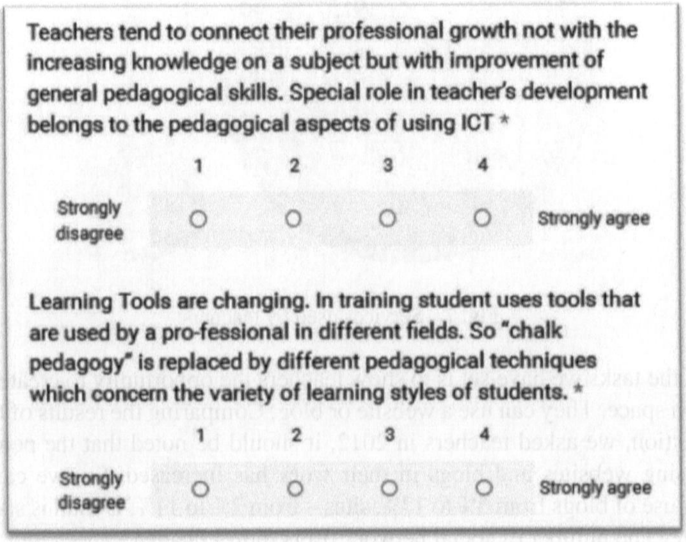

Fig. 9. A fragment of the third block survey

It was not a set of questions but most likely thesis with which most teachers agree. In fact, all the questions received a response "strongly agree." The unanimous approval was the fourth point. Here is a list of thesis.

1. Teachers tend to connect their professional growth not with the increasing knowledge on a subject but with improvement of general pedagogical skills. Special role in teacher's development belongs to the pedagogical aspects of using ICT the variety of learning styles of students.
2. Learning Tools are changing. In training student uses tools that are used by a professional in different fields. So "chalk pedagogy" is replaced by different pedagogical techniques which concern.
3. Teacher's mark is increasingly replaced by students' self-esteem, their mutual and automatic estimation as well. The function of a mark is not to evaluate students perform on the tasks but target students on what is needed to be done to extend academic work and achievement.
4. The teacher does not know and should not know all the answers. Students do not expect that the teacher gives them the answers to the questions, but they hope they get help in finding the answers by themselves. Answers to the questions can be found using many different sources, including the Internet.

5. The educational space expands beyond the classroom. The teacher encourages students to study in various environments, including outside schools both in real and virtual environments. Much of the work takes place in a real world, with the participation of fellow student and other adults other than teachers.
6. Individual work becomes a common activity. Most of the students do academic work together with classmates or within the creative (research) groups, using a computer and the Internet as the main tool of their activities. The teacher is preparing educational materials in digital format, exchanging designs with colleagues inside and outside the school.
7. Expanding the range of knowledge and educational information. The student determines and is an essential source of information and resources. The teacher encourages the materials and tools that can be used by student. More and more teacher start to use examples from real life, and not limited to the scope of textbooks only.
8. Increasing of students' independence and responsibility for the results of educational activities. The teacher teaches only the core disciplines form the students' ability to develop the subject beyond this core independently. Student determines what to do, based on existing knowledge and an idea of what is needed to know. The role of teachers is reduced to a more general management in different types of work that initiate and carry out the students themselves.

Survey results indicate that the professional development of teachers is, in particular, develop the ability to find information, analyze it and implement in practice, to respond quickly to innovative experiences, as well as to design, create, experiment innovations and be able to distribute purposefully.

The lowest asset was given to the 3d point which is connected with the change of the role of evaluation at school. Teachers' acquaintance with molding evaluation was one of the goals of the "Intel@Teach to the Future" program. Self-esteem – is one of the priorities skills for successful professional development and lifelong learning. Minor teachers' support of this idea is the basis for us to develop this direction in the future (i.e. to add in courses thematic blocks and tasks aimed at development of formative assessment skills, to acquaint teachers with the appropriate on-line services on master classes). Low estimate was given to point 1 and 5 which is also directly related to the work of teachers. However, unanimous approval was given to point 4, which increases student's responsibility for learning achievements.

Important is the desire expressed by teachers in more detailed use of new trends in education in their work (Fig. 10).

In the last survey, respondents noted what brands and services they would like to explore in more detail and use in their work. All of the teachers indicated that they wanted to use those services in the future which they learnt on a master class. The lowest support was given to STEM-education and Augmented Reality. In our opinion, it is connected with the material investment that needs to be done to implement these trends in education.

The high level of support inverted class technology and gamification were received. Tools and Web services for these technologies (interactive video EDpuzzle, LearningApps) are the most popularity among teachers due to their ease of development and a broad range of applications.

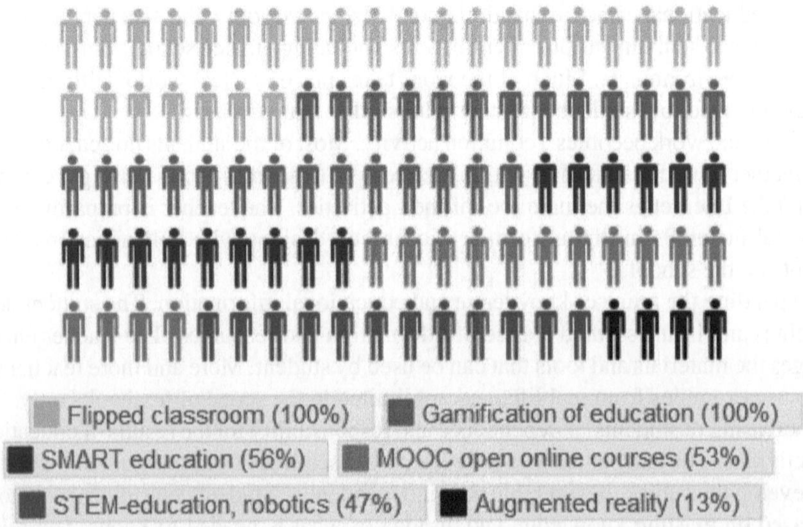

Fig. 10. The level of interest in the development and use of new educational trends

The main problems were different ICT-level training participants of the master class, in particular, slow registration on web-services, access to the account (mail, social network, etc.).

7 Conclusions and Outlook

Technology of master classes can be a good alternative to long-term courses. This approach is well accepted by teachers and has several advantages like: short term, high work productivity, creation of teaching materials using web services, ready to use in practical activity. During the master class the main goal becomes not to teach how to own the technology but to show its advantages and persuade to use a particular technology for pupils' learning, to demonstrate the ease of its use, motivate teachers to further development of new services and the their use in the professional activity.

An important result is shifting teachers' point of view to changes in the education system under the influence of digital technology, learning and adopting new trends.

References

1. Arhipova, T., Osipova, N., Lvov, M.: Social networks as a means of learning process. Inf. Technol. Educ. **22**, 7–18 (2015). (in Ukrainian)
2. Baran, E.: A review of research on mobile learning in teacher education. Educ. Technol. Soc. **17**(4), 17–32 (2014)
3. Berk, R.A.: Teaching strategies for the net generation. Transform. Dialogues Teach. Learn. J. **3**(2), 1–23 (2009)

4. For half of the Ukrainian Internet use; two-thirds of them - in social networks. http://www.gfk.com/uk-ua/rishennja/press-release/vzhe-polovina-ukrajinciv-koristujutsja-internetom-dvi-tretini-z-nikh-v-socialnikh-merezhakh. (in Ukrainian)
5. Kherson Virtual University. http://dls.ksu.kherson.ua/
6. KSUOnline. http://ksuonline.kspu.edu/
7. Kushnir, N., Manzhula, A., Valko, N.: Future and experienced teachers should collaborate on ICT integration. In: Ermolayev, V., Mayr, H., Nikitchenko, M., Spivakovsky, A., Zholtkevych, G. (eds.) ICTERI 2014. CCIS, vol. 469, pp. 217–237. Springer, Cham (2014). doi:10.1007/978-3-319-13206-8_11
8. Kushnir, N., Manzhula, A.: Formation of digital competence of future teachers of elementary school. In: Ermolayev, V., Mayr, H.C., Nikitchenko, M., Spivakovsky, A., Zholtkevych, G. (eds.) ICTERI 2012. CCIS, vol. 347, pp. 230–243. Springer, Heidelberg (2013). doi: 10.1007/978-3-642-35737-4_14
9. Morze, N.V.: How to train teachers to computer technologies ceased to be a miracle in training? Comput. Sch. Fam. **6**, 10–14 (2010). (in Ukranian)
10. Seppälä, P., Alamäki, H.: Mobile learning in teacher training. J. Comput. Assist. Learn. **19**(3), 330–335 (2003)
11. Shovkun, V.: Training of future teacher of informatics to work in modern information and educational environment of school. Inf. Technol. Educ. **23**, 136–146 (2015). (in Ukrainian)
12. Soldatova, G., Zotova, E., Lebesheva, M., Shlyapnikov, V.: Digital Literacy and Internet Security. Methodical Manual for Specialists Basic General Education. Google, Moscow (2013). (in Russian)
13. Spivakovska, E., Osipova, N., Vinnik, M., Tarasich, Y.: Information competence of university students in Ukraine: development status and prospects. In: Ermolayev, V., Mayr, H.C., Nikitchenko, M., Spivakovsky, A., Zholtkevych, G. (eds.) ICTERI 2014. CCIS, vol. 469, pp. 194–216. Springer, Heidelberg (2014). doi:10.1007/978-3-319-13206-8_10
14. Spivakovskyi, A., Vinnik, M., Tarasich, Y.: Using ICT in Training Scientific Personnel in Ukraine: Status and Perspectives (2015)
15. UNESCO Policy Guidelines for Mobile Learning. http://unesdoc.unesco.org/images/0021/002196/219641E.pdf
16. Unwin, T.: Towards a framework for the use of ICT in teacher training in Africa. Open Learn. J. Open Distance e Learn. **20**(2), 113–129 (2005)

Author Index

MIX
Papier aus verantwortungsvollen Quellen
Paper from responsible sources
FSC® C105338

If you have any concerns about our products,
you can contact us on
ProductSafety@springernature.com

In case Publisher is established outside the EU,
the EU authorized representative is:
Springer Nature Customer Service Center GmbH
Europaplatz 3, 69115 Heidelberg, Germany

Printed by Libri Plureos GmbH
in Hamburg, Germany